Creativity and the Brain

Editors

Mario Tokoro
Ken Mogi

Sony Computer Science Laboratories Inc., Japan

Creativity and the Brain

World Scientific

NEW JERSEY • LONDON • SINGAPORE • BEIJING • SHANGHAI • HONG KONG • TAIPEI • CHENNAI

Published by

World Scientific Publishing Co. Pte. Ltd.

5 Toh Tuck Link, Singapore 596224

USA office: 27 Warren Street, Suite 401-402, Hackensack, NJ 07601

UK office: 57 Shelton Street, Covent Garden, London WC2H 9HE

British Library Cataloguing-in-Publication Data
A catalogue record for this book is available from the British Library.

CREATIVITY AND THE BRAIN

ISBN-13 978-981-270-019-3 (pbk)
ISBN-10 981-270-019-6 (pbk)

Typeset by Stallion Press
Email: enquiries@stallionpress.com

Printed by FuIsland Offset Printing (S) Pte Ltd, Singapore

Preface

Until recently, human creativity had been treated as a mysterious process of brain activity, since we had neither the tools to measure brain activity precisely nor the theories to analyze and synthesize creativity. However, the quick advancement of brain science these days — thanks largely to the development of various measurement tools such as electroencephalography (EEG), electromyography (EMG), and functional magnetic resonance imaging (fMRI), and to new attitudes which view the brain as a system — has enabled us to discuss creativity in the context of science. Hence, we organized a workshop for the exchange of ideas among distinguished scholars who have been contributing to the understanding of creativity in different fields and from different viewpoints.

The workshop was held from April 27–30, 2004, in an ancient castle located in a small village called Bertinoro, near Bologna, Italy. We invited the following distinguished scientists from around the world: Professor Nobuo Masataka (Kyoto University, Japan), Professor Ernst Poeppel (Ludwig-Maximilians University, Germany), Professor Philippe Rochat (Emory University, USA), and Professor Allan Snyder (Director, Centre for the Mind, Australia). We also asked the following researchers from Sony Computer Science Laboratories (Sony CSL) to participate: Dr Hiroaki Kitano (Director, Sony CSL, Tokyo), Dr Takahiro Sasaki (Researcher, Sony CSL, Tokyo), and Dr Luc Steels (Director, Sony CSL, Paris).

As expected, many brilliant and ingenious ideas were presented at the workshop, and we felt that it was important to record and share these ideas with those who are interested in this area. Thus, we decided to publish a book based on the discussions we conducted in the workshop. We also decided to publish the book not as a conference report, but as scientific literature to widen the audience while simultaneously maintaining scientific depth. For these purposes, we asked a science writer, Dr Kaoru Takeuchi, to participate in the workshop and in our editorial work. He has provided a short introductory note for each lecture/chapter.

This book is composed of eight chapters. In chapter 1, titled "A Genius Within," Professor Allan Snyder, a brain scientist and multitalented genius, presents the notion of "mindset." A mindset is composed of an individual's past experiences, so that most of his or her decision making and actions can be done in a routine manner. Snyder claims that creativity lies in breaking the routine. He has built a device that emits very weak electromagnetic pulses to the left temporal lobe in order to mask one's mindset to show how creativity emerges.

In chapter 2, titled "A Story of Brain Clocks," Professor Ernst Poeppel, a neurophysiologist, shows that the temporal resolution of causal relation in the human brain cannot be shorter than 30 milliseconds. This is the resolution of the perceptual process. He also shows that the brain can remember or perceive events that last less than 3 seconds. This is called the chunk of conceptual process. These numbers come from the brain's physical structure. He hypothesizes that creativity lies in the relationship or complementarity between the perceptual and conceptual processes.

In chapter 3, titled "In Search of Achilles' Heel," Dr Hiroaki Kitano, an advocate of systems biology, explains that interesting phenomena occur on the border of stability and instability, and that a complex system incurs fragility in order to be robust. He cites diseases such as cancer and diabetes as examples, but the discussion can also be applied to the brain system and creativity.

In chapter 4, titled "Do Infants Dream of Baby Sheep?" Professor Philippe Rochat, an expert on the developmental psychology of infants, explains the developmental stages of infants' brain. Babies less than 2 months of age are similar to prenatal babies. Then, babies begin to realize that their bodies belong to them. After 9 months, they become somewhat social. A human being is a social animal; however, Rochat claims that creativity emerges when people stop worrying about what other people think.

In chapter 5, titled "Baby Talk," Dr Nobuo Masataka, an expert on the communication of primates, talks about physical development and its relation to expressional ability, based on his study of apes. He also discusses the problem of learning words. He points out some common elements in the learning of humans and apes based on social relations and interactions with the environment, culminating in human language.

In chapter 6, titled "Dr Jekyl, Mr Hyde, and Qualia" Ken Mogi, an advocate of systems brain science, shows that every cognitive process is necessarily metacognitive, using the homunculus as a means for explanation. He argues that creativity is an inherent property of the brain system, in view of the way the brain realizes the coexistence of dynamical adaptability and perceptual stability, the latter correlating with the qualia in conscious experience.

In chapter 7, titled "Computer and Creativity," Luc Steels, a computer scientist and an expert on complex systems, claims that creativity emerges from the interaction with others. He believes it is important for a group of people or for society as a whole to be creative. He also claims that the expression of meaning is creative, rather than a realistic drawing.

Chapter 8, titled "Dialog by Editors," is a sort of addendum to this book in order to help readers re-experience the atmosphere of the workshop and deepen their understanding of the lectures. Mario Tokoro and Ken Mogi, with the help of Dr Takahiro Sasaki, present discussions on each lecture. They also discuss the relations between the lectures, so that the whole picture of the workshop can be easily drawn.

This book is part of a long-term effort, initiated by Mario Tokoro, to weave a web of researchers and practitioners that are dedicated to advance the quality of education in the 21st century. The initiative intends to stimulate multidisciplinary discussions, which contribute to improving education and learning for children and adults of all ages everywhere in the world, based on an improved understanding, methodology, and policy for education and human learning.

This book follows up on two previously published books. The first book, titled *The Future of Learning: Issues and Prospects* and published by IOS Press in 2003, sets the scene for our discussion. The second book, titled *A Learning Zone of One's Own* and also published by IOS Press in 2004, discusses in-depth issues concerned with grounding, optimal experience and emotion, and pedagogical methodologies. We hope that the current book on creativity and the brain, together with the other two books, provides a whole new aspect and sheds light on the problem of learning and creativity in the cognitive and brain sciences as well as education.

We are much indebted to all the distinguished scientists who participated in the workshop. We thank Dr Takahiro Sasaki of Sony CSL, Tokyo, for his valuable suggestions in organizing the workshop and in editing this book. We also thank Ms Yumiko Kitamori for her help in organizing the workshop. We are grateful to Miss Ho Sheo Be, Senior Editor, World Scientific Publishing Company, for her effort in preparing the final form of this book.

Tokyo, October 2005

Mario Tokoro, President,
Sony Computer Science Laboratories, Inc.
Ken Mogi, Senior Researcher,
Sony Computer Science Laboratories, Inc.

Contents

List of Contributors

Allan Snyder
Centre for the Mind
University of Sydney
Sydney, NSW 2006
AUSTRALIA

Ernst Poeppel
Ludwig-Maximilians University
Goethestrasse 31
Munich D-80336
GERMANY

Hiroaki Kitano
Sony Computer Science Laboratories, Inc.
Takanawa Muse Building
3-14-13 Higashigotanda, Shinagawa-ku
Tokyo 141-0022
JAPAN

Philippe Rochat
Emory Infant Laboratory
318 Psychology Building
532 Kilgo Circle
Emory University
Atlanta, GA 30322
USA

Nobuo Masataka
Cognition and Learning Section
Primate Research Institute
Kyoto University
41 Kanrin, Inuyama
Aichi 484-8506
JAPAN

Ken Mogi
Sony Computer Science Laboratories, Inc.
Takanawa Muse Building
3-14-13 Higashigotanda, Shinagawa-ku
Tokyo 141-0022
JAPAN

Luc Steels
Artifical Intelligence Laboratory
Vrije Universiteit Brussel and
Sony Computer Science Laboratory Paris
6 rue Amyot
75005 Paris
FRANCE

Mario Tokoro
Sony Computer Science Laboratories, Inc.
Takanawa Muse Building
3-14-13 Higashigotanda, Shinagawa-ku
Tokyo 141-0022
JAPAN

Takahiro Sasaki
Sony Computer Science Laboratories, Inc.
Takanawa Muse Building
3-14-13 Higashigotanda, Shinagawa-ku
Tokyo 141-0022
JAPAN

Kaoru Takeuchi
1-5-3-313 Jomyoji Kamakura
Kanagawa 248-0003
JAPAN

A Genius Within

Tall, lean, and mildly eccentric, Professor Allan Snyder never takes off his black leather cap, even inside a restaurant. He is a professor from Australia, a Fellow of the Royal Society of London, who smiles like a wizard and speaks like a wild New York teenager.

He has a surprising present for you — a wonderful gift, indeed. If you visit his laboratory, he will put an electromagnetic brain-stimulating device on your head that will turn you into a genius. This is the magic of the 21st century.

Sounds too good to be true? Well, some people say that ideas in science must be crazy enough to be true.

The gist of Professor Snyder's idea can be briefly summarized as follows:

1. *Autistic savants without any learning skills sometimes show amazing talent.*
2. *This is because autistic savants somehow have direct access to raw or lower-level algorithms of the brain.*
3. *Healthy adults, on the other hand, are denied this access.*
4. *By electromagnetically stimulating the left side of the brain, however, healthy adults might be able to gain this access.*
5. *In a word, you have a "genius" hiding inside your brain.*

So, Professor Snyder claims that autistic savants reveal the secret of how our inner brain works, and that we can mimic this savant effect by electromagnetic stimulation.

Some day in the near future, you might be able to experience the world of a genius via a portable "genius machine," or what Professor Snyder calls a "thinking cap."

Learning and Creativity — Accelerated by Suppressing or Circumventing Certain Brain Regions

Allan Snyder is the director of the Centre for the Mind, a joint venture between two prestigious Australian universities, the Australian National University and the University of Sydney. He holds the 150th Anniversary Chair of Science and the Mind at the University of Sydney, and the Peter Karmel Chair of Science and the Mind at the Australian National University. He is also a fellow of the Royal Society of London.

Allan Snyder is a graduate of the Massachusetts Institute of Technology (MIT), Harvard University, and University College London. Previously, he was a Guggenheim fellow at the Yale University School of Medicine and a Royal Society guest research fellow at the Physiological Laboratory, Cambridge University, England.

Allan Snyder received the 2001 Marconi International Prize — "the world's foremost prize in communications" — in New York City, and the 2001 Clifford Patterson Prize of the Royal Society in London.

Champion Mindsets

Let me begin by putting learning into perspective. Ultimately, I care about learning to be creative, to make leaps. I make the distinction between holding something up and saying "learn that," and being truly creative. We care about learning to be creative. After all, any company or corporation — I'm particularly interested in corporate creativity — is only as good as its next big idea; and creativity is the driving force that incubates these big ideas.

So, we need scientific ways to identify creative individuals in order to assemble innovation teams, for example. At the Centre for the Mind, we have developed a unique measure called the "creativity quotient" (CQ), which generates a creativity profile for each and every person in a corporation. However, I'm not going to talk about that today; I'm just bringing it up as background.

We also need strategies to nurture creativity, starting from the earliest of ages. We have developed a program at the Centre called "Any Kid can be a Champion" that does just that, but I'm not going to talk about that today either. Instead, I will talk briefly about my book *What Makes a Champion!*[a] during this lecture.

A person's creativity does not depend on how fast or how much he or she learns. Indeed, studies have shown that child prodigies, who learn faster than anybody else, rarely amount to anything (although there are some exceptions, like Mozart). On the other hand, all of the Nobel laureates and fellows of the Royal Society of London — I just happen to have my hands on that because I am a fellow — scored very average grades in school. This is astonishing: Nobel laureates and fellows of the Royal Society of London were very average students. In other words, if your children are doing great in school, then they will probably not become brilliant scientists.

This is nothing to agree about. This is a fact. I haven't said anything interesting. I'm just talking about facts. I really mean it.

What is really important is learning how to struggle, how to recover from adversity, and how to adapt. Nothing is more important than daring to take risks and to confront conventional wisdom. If you just learn automatically like a prodigy does, you will never learn how to struggle.

[a] Snyder A (ed.), *What Makes a Champion!* Penguin, Australia, 2002.

I was a lousy student, and I was ashamed of it (although I have to admit that I wasn't so lousy that I could not get into Harvard). I think things came hard to me, so I had to learn how to struggle. This is what I call the "champion mindset": being able and willing to take risks, confront conventional wisdom, and face adversity.

What Makes a Champion! deals precisely with this concept. I arranged for Nelson Mandela and a group of 50–60 extraordinary people to be together, and we had a 3-day conference about this. What I discovered was that creativity is an act of rebellion. To be creative, you have to confront conventional wisdom; you have to break with convention. If you are not willing to be subversive, you will not be creative. After all, something is considered a breakthrough only when it is subversive.

The hallmark of someone who has this champion mindset, as I call it, is that they abhor being just average or ordinary. Now, that's surprising. When I started out, I thought that to be extraordinary or truly creative, one would be driven by the need to be successful or by the fear of being a failure. However, this is not the case. Champions are not by nature driven to succeed, neither are they driven by a fear of failure; they simply do not want to be ordinary. They want to differentiate themselves from the rest of the pack, and not always consciously. Becoming a champion is not a conscious decision that you can simply choose to do: if it is conscious, then you have probably destroyed any hope of ever achieving it.

So, my take-home message on this little preamble is that we should not encourage children to be like everyone else (this message really drives home the debate over Asian versus Western learning). Instead, we should nurture their uniqueness. If you want children to have this special ingredient that we call supercreativity, then you have to encourage them to be subversive in order to convert conventional wisdom.

Now, there are a lot of families and individuals that I could talk about and use as examples, but I'm going to just briefly talk about me and my family. I take no credit for anything about me — my parents, as you will see, must take credit, and in a extraordinary way. I come from a family of three boys, and each of us has won major international prizes in different fields. I have no idea why this is so. I can, however, tell you that my parents constantly told us to nurture our uniqueness: "Each of you has something to give, whatever it may be," they would say.

You are all unique. By definition, no one else is like you, and so each of you has something that is different from everyone else. Therefore, if you develop that, then you will, by definition, have this extraordinary thing. So wham! Celebrate that!

Mindsets

Now, I'd like to talk about something that was the subject of an article I wrote in *Nature*.[b]

In my opinion, the major challenge for the future is to have devices that help amplify a person's creativity. I've told you what I think creativity is, but I'm more interested in constructing devices that amplify a person's creative learning ability: devices that allow you to realize your true genius, your true creative capacities. Ultimately, my dream is to have a learning machine — a real "thinking cap."

The bottleneck to creative learning — as opposed to just learning, which you usually do in school (e.g. rote memory) — is failing to recognize that there is something new to learn in the first place. The bottleneck to learning is not that people have trouble learning (I've already told you that prodigies don't amount to anything), but rather that most people look at something and don't recognize there's something to learn. This is profoundly important, and is backed up by neuroscience and psychology. In fact, I'm going to back it up right now.

Unlike infants, we as adults do not see the world through a literal interpretation of our senses. Instead, we see it through our "mindsets": our knowledge which has been accumulated through past experiences.

[b] Snyder AW, Barlow H, "Human vision: revealing the artist's touch," *Nature* **31**: 117–118, 1988.

Now think about that. How can we join the dots up in a different way if, every time you look at the dots, you impose something you already know? I like this metaphor because I don't think there's any better way other than this beautiful metaphor to describe all the information I'm about to give you.

Our brains impose meaning on sensory information. They present us with a coherent picture or the best hypothesis, when in reality there is always ambiguity. In other words, we are designed to be experts. We are wired to maneuver rapidly in a familiar environment. However, this comes at the cost of prejudice, the cost of making assumptions. It is a tremendous evolutionary advantage to have mindsets or concepts, but they incur large costs when it comes to making breakthroughs.

Compelling evidence comes from illusions. For example, we all make an implicit presumption that light comes from above. Our visual system is wired to assume that light comes from above. If light comes from below, we make terrible mistakes, such as concave turning to convex.

As another example, we often read what we know, not what is really there. Read this:

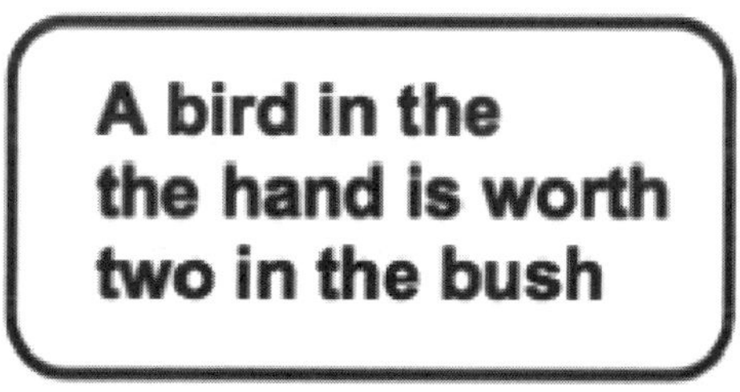

Most people who read this miss the double "the." Oliver Sacks read it 25 times in my office, and got so angry that he left the room! You read what you know.

Imagine two people looking at the same cloud formation. The portrait painter sees a face of dignity, whereas the ultrasound sonographer sees a diseased gallbladder. You see what you know. As yet, the cloud is neither a portrait nor a diseased gallbladder, but you see what you know.

Look at this:

Most people in time see a Dalmatian. Now, how could you ever see a dog if you didn't already have a mindset for Dalmatians? You cannot tell that there is a dog there unless you already have a template or mindset for a dog.

• • •

Debate

Steels: Is it a dog or a structure?

Snyder: Well, most people see a Dalmatian.

Steels: Yes, I know what your point is. You cannot see a dog if you didn't have the mindset of a dog. What I'm saying is that it must be a dog — there has to be some kind of coherence that gives rise to the concept.

Snyder: That's an interesting concept. When you look at the cloud and you see a portrait, is there some coherence in the portrait? I claim that there isn't. In fact, the way I see it, you walk around the world projecting what you know on everything.

Steels: I don't agree.

Snyder: Oh good, good. Let me try to go on, yeah. I'm glad you don't agree. If I were preaching something that you agreed with, what good would that be!

• • •

Now, the interesting thing is that when I showed this same diagram to a South African person, he didn't see a Dalmatian. Instead, he saw a hyena — a different animal — looking in a different direction. So, you see what you know. The South African was familiar with hyenas, so he saw a hyena; he never saw a Dalmatian. He had great trouble shifting his interpretation from a hyena to a Dalmatian. In fact, I had to show him pictures of real Dalmatians to get him to change his mind.

There's no doubt that we see what we know, and that we are often blind to alternative interpretations. We are blinded by our mindsets. We are prejudiced to know our life by our past experiences. This is extremely difficult to overcome.

Over the years, I've shown this picture to thousands of people, with very interesting outcomes. Each person sees a different thing, depending on the person's background; ornithologists see birds, for instance. I cannot think of anybody who has offered more than one interpretation.

The story is even more intriguing. Recent evidence in *Science* shows that the sensory system itself falsifies the signal. In other words, when the executive brain gets an idea, it commands the lower levels to falsify the idea to conform to the executive decision. The executive brain subconsciously makes an interpretation of something and, before you know it, the sensory information is demanded to conform to it (I can give you the references, but it's not my work). So, amazingly, the mind tells the lower sensory levels to lie. I think this is profound.

So, what does this mean? I'm arguing that this is a consequence of the brain having mindsets. Mindsets are good things, by the way; don't get me wrong. They allow us to maneuver fast and be experts for most of our lives. If you have a stomachache and go to the doctor, the doctor could be right in saying, "Ah, we have to be very careful about this. It could be aliens running around in your stomach." However, when it comes to creative science, this is not good.

What are mindsets? They are collections of sensory details that characterize familiar objects, or objects in the most general sense. Mindsets are embodiments of the familiar.

So, to sum up: our brains see the world through mindsets — mental templates, if you like, of the world derived from past experiences. While templates are very good expertise, they also blind us to novelty. Therefore, the bottleneck to learning is actually failing to recognize that something is new. We see what we know; we see only the whole, not the parts.

What is the neurobiological mechanism that prevents us from seeing the world literally? Well, prior to the formation of a mindset, while we are still in infancy, I believe that we are aware of the lower-level sensory information pouring into our brains. However, once a mindset or concept is formed, whether it is learning how to drive a car or something else, we are no longer aware of the lower-level sensory information.

This is just a simple conceptual model showing that you are aware of all the details before you have a mindset:

Lower-level sensory information
↓
Executive (conscious) awareness

However, once you have developed a mindset or concept, you see only the label. You see the big picture, not the details.

Lower-level sensory information
↓
Concepts
↓
Executive (conscious) awareness

For example, we are not consciously aware of how our brains derive the shape of an object from the subtle shading around its surface. If I give you a ball and ask you to draw a ball, you will not be able to draw a ball unless you have prior experience or teaching. Because the ball is the ball, because of the subtle shading around the ball, and because your brain is an expert about seeing, you do not see the details. Your brain just says as it should: "Ball." You are aware of the ball as a whole (i.e. on the level of the label), but not in terms of the details that attribute to make up the label.

The Best of Both Worlds? Case Studies of Autistic Children

How can we circumvent our mindsets when the concept is an intrinsic part of our makeup? Surprisingly, it takes insights from damaged brains — abnormal minds — to answer this question.

Curiously, some brain-impaired individuals who suffer from infantile autism are the very opposite of us. They see only the parts, and do not tend to see the whole.

Let me remind you what our ultimate goal here is: we want to make a creativity machine that can accelerate creative learning. To do this, we must — at least momentarily — be literal, free from prior interpretations. Thus, I'm looking at autistic children to try to understand why they can see the world literally. In other words, could we have the best of both worlds? Could we have our mindsets for one moment, and then switch them off the next to see the dots as they are, to see the world literally?

I'm trying to make a creativity machine. I'm trying to make you, at least momentarily, literal so that you can see the world the way it is. You can see everything without prejudice, so to speak.

How can you do that? Well, there are several interesting ways. One is by trickery. For example, if you want to draw, you can get rid of the meaning of what you see by turning the drawing upside down or by looking at the drawing one piece at a time. However, tricks like these have their limitations.

Another way to bypass mindsets is by brute force. This involves turning the mindset off by turning off part of the brain. A perfectly safe way of doing this is via transcranial magnetic stimulation (TMS), in which the left temporal lobe is shut off. Why the left temporal lobe? Because autistic children tend to have damaged left temporal lobes, especially the ones that are literal.

So, I mimic the behavior of autistic children by turning off the left temporal lobe with magnetic pulses. Magnetic pulses are known to shut off local parts of the brain.

Here is a normal 4-year-old child (with an IQ of 40) who draws like caricature, drawing not what is seen, but what is known.

Representative drawings of normal children, each at age 4 years 2 months (Emma and Teneal, parents on campus preschool, Australian National University).

Here is an autistic 3-year-old girl who, by the way, cannot recognize her own mother and the nurse, cannot tie her own shoelaces, cannot speak a single language, and never drew before in her life; yet, she starts drawing literally what she sees. She draws, to me, like the master himself, Da Vinci (who spent years mastering the techniques of drawing). This really challenges the notion that certain things take years of experience, that we obviously see things based on mere mimicry of what is already in our brain.

Autistic child's drawing at about 3½ years (Selfe 1977)[c]

I've also taken perfectly normal 20-year-old students to see how they draw both before and after TMS. Before I gave them magnetic stimulation, they drew what I would expect: outlines. Then, after 10 minutes of magnetic pulse stimulation to the left temporal lobes, they were much more realistic. Forty-five minutes later, they went back to normal.

---

[c] Selfe L, *Nadia: A Case of Extraordinary Drawing Ability in an Autistic Child*, Academic Press, London, 1977.

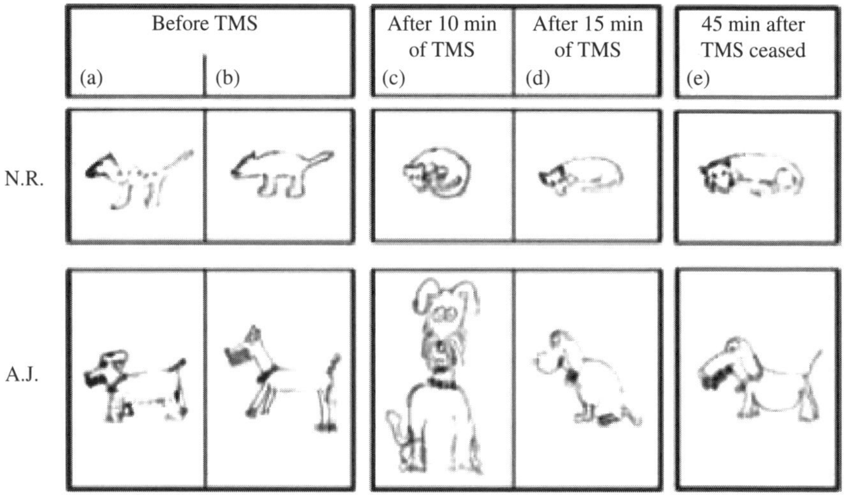

Effect of transcranial magnetic stimulation (TMS) on drawing ability in two participants. The figure illustrates a dog drawn from memory by participants N.R. and A.J. (a, b) Practice session before the application of TMS; (c) during TMS; (d) immediately after TMS; and (e) 45 minutes after TMS ceased.

After undergoing TMS, the students completely changed their schema, their way of looking at life. They became literal. This is quite amazing.

• • •

Discussion

Poeppel: About the 3-year-old artistic girl, did you confirm the presence of impairment in her left temporal lobe?

Snyder: Yes, she had left temporal thin lobe impairment.

Poeppel: Have you tried your method on yourself?

Snyder: Ah yes, I have.

Poeppel: And what was the result?

Snyder: Oh look, I'm a bad subject because I know what I want, and I can't block out knowing what I want. I perform well, but no one wants to believe that I do because I wanted it to work to begin with. So, I'm not a good subject.

Poeppel: What did you experience, for example, in terms of colors and senses?

Snyder: I didn't experience colors, but I had a more conscious awareness of detail. It's subtle. This is not a profound switch. It works in about 45% to 60% of our test subjects. Why it doesn't work in the rest, no one knows.

• • •

Thinking Cap: A Perspective

At the Centre, we have made people literal by using these procedures. I think the world should be interested in learning the ways to circumvent mindsets. For instance, you can engage a person in electroencephalogram (EEG)-assisted neurofeedback by identifying the literal, nonconscious brain states.

I'm not going to tell you exactly how the Centre identifies these states because that is confidential. I can, however, tell you two general conceptual procedures we have used to do this. One way is by looking at the EEG of autistic students or individuals, who are usually literal when doing tasks. The other way is to look at the EEG of normal people when they become literal after being magnetically stimulated. Through these two types of procedures, you can identify literal brain states.

In conclusion, I emphasize that the bottleneck to creative learning is, surprisingly, not knowing what to learn. The bottleneck to creative learning is actually not being aware of what is new. We need — just for a moment — to be literal, to see the dots the way they are. This can be achieved via EEG neurofeedback using a "thinking cap," which can be wireless and which interacts with both the computer and the person wearing the headband.

I see this as the next big technological leap that will accelerate creative learning. It is a totally unique way of seeing the world differently. Among the many devices I have thought up, the "thinking cap" is tailor-made for this particular learning workshop. Thank you.

Further Reading

Snyder A (ed.), *What Makes a Champion!* Penguin, Australia, 2002.

A Story of Brain Clocks

Imagine playing one of Beethoven's piano sonatas. A good pianist knows how to keep the tempo. But how? How is it possible, biologically, to control the rhythm? Is there a biological metronome inside his or her body?

The answer is yes, according to the neurophysiologist Professor Ernst Poeppel. It is not, however, as you might have guessed, your heartbeat that regulates your musical tempo. Surprisingly, a tiny metronome resides inside your brain.

Before going into Poeppel's hypotheses, let me briefly review the basics of brain physiology (or psychology).

Let's perform an experiment as shown in this figure. You are the subject. You hear two simultaneous clicks. Interestingly, you feel as if you've heard only one click; you cannot distinguish the two. Moreover, this sound is not felt in the center of your brain, but is shifted a little bit to the left of your brain. Why?

Well, this is because the brain's left hemisphere is better at catching (and analyzing) sounds than the right hemisphere. When you hear a sound, your left brain is in control.

Next, you are exposed to two distinct clicks with a time interval of, say, 1 millisecond (1/1000th of a second). This time, you know there is a time delay, but you cannot tell which sound came first. The interval is too short for your brain to tell the "before–after" relation.

When the time interval becomes approximately 30 milliseconds, you may say something like, "I heard the left click, then a bit later came the right click." In other words, in order for your brain to identify two distinct sounds, they must be temporarily separated by at least 30 milliseconds.

Now, let's turn to visual senses. Again, you are exposed to two distinct stimuli. This time, you see two successive lights in front of you. Can you tell that there are two light pulses?

Unlike your ears, your eyes cannot distinguish two pulses when their time interval is less than 10 milliseconds. Your eyes are "slow" compared with your ears. This is because, inside your brain, the information of light must be chemically transformed before being transmitted. Olympic champions in ping pong, for example, must have good ears rather than good eyes. Light travels fast in the air, but slows down significantly once it gets into your brain. Sound waves, on the other hand, travel slow in the air, but move pretty fast inside your brain.

What is important here is to realize that the notion of time is not simple. Forget about physics; we do not need Newton or Einstein here. Our brain has its own temporality. What is simultaneous in physics is not necessarily simultaneous brainwise. The same can be said of the "before–after" relation.

Whatever the modality (whether the senses are visual or auditory in nature), in order for your brain to identify and tell which information came first, the information must be separated by at least 30 milliseconds. This is the magic number.

Professor Poeppel's other interesting magic number is 3 seconds. According to this astonishing hypothesis, the human experience of "now" is not instantaneous, but has the temporal duration of nearly 3 seconds.

How is this possible? Can we prove this somewhat bizarre statement? Well, just look at the following picture. This is called the Necker cube.

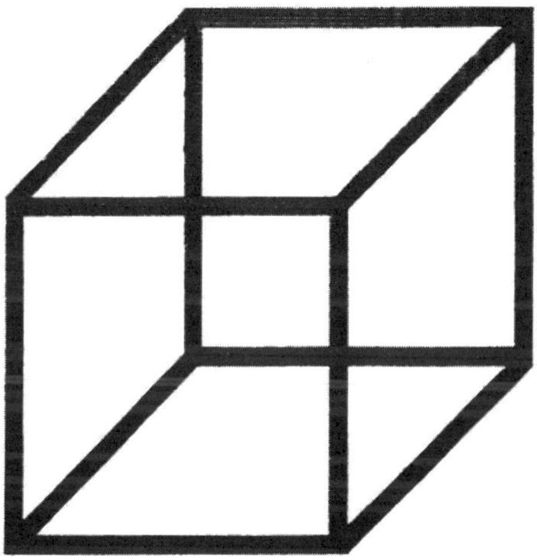

Did you notice that there are two different cubes here? You experience the switch in perspective — every 3 seconds or so. Try hard to prevent this switching. Can you do it? Please try for a minute.

You probably could not prevent the switching. Actually, nobody can stop the switching, which occurs almost every 3 seconds. For the sake of clarity, 3 seconds is not a constant of physics; it varies slightly from person to person. Nevertheless, on average, people "refresh" their window of consciousness every 3 seconds.

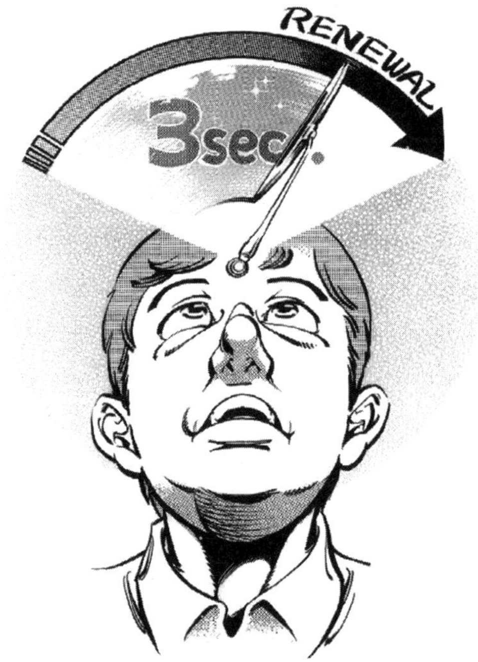

This can be likened to a computer screen. Usually, a computer screen is renewed every 10 milliseconds. Your consciousness, on the other hand, is renewed every 3 seconds.

Complementarity and Creativity in Brain Sciences

Ernst Poeppel is a professor of medical psychology (Ordinarius) and the chairman of the Institute of Medical Psychology, both at the Ludwig–Maximilians University in Munich.

He studied psychology and zoophysiology at the University of Freiburg in Breisgau, the University of Munich, and the University of Innsbruck; and received a PhD in psychology from the University of Innsbruck.

He is a member of the European Academy of Sciences and Arts; and is a recipient of many awards including the 1988 J. E. Purkinje Medal of the Czech Medical Societies, and the 1984 Levinson Award of the American Poetry Association.

Chance and Meaning

Good morning. Thank you very much for inviting me.

I was born in what is now Poland in 1940. Six years later, my family and I were put in a train, and we were transported from east to west. Our train traveled 200 km farther than the train my extended family was in. So, I grew up in West Germany, while the rest of my family grew up in the East. One of my cousins (with whom I lost touch) made a political career for himself in East Germany, and became a well-known member of the ruling communist party in the German Democratic Republic. It was just by accident, a chance event, that my train went further west. There are a lot of events like this; I think all of our lives are chance events. But, what is interesting about it is that we immediately create meaning.

I think the complementarity of chance and meaning is very fundamental for our lives. For example, we meet here at this workshop. But why? Whatever the answer is, it is a retrospective construction of meaning. So, I think this is complementarity: chance and meaning. This concept is important for understanding creativity. We create meaning out of something that has just happened.

In my talk, I want to first make a few general remarks about evolution, which demonstrates the creativity of nature around us. Then, I will tell you about conceptual brain science — how this brain machinery, which is the locus of creativity, works. Finally, I want to make a few remarks on what is necessary to develop a taxonomy of creativity; namely, the variables to consider in such a taxonomy.

Before I begin, I should tell you that I have prepared handwritten transparencies for my talk because I cannot think when I type. This is another point I want to make: the mode of expression is essential for creativity. Creativity is not abstract, but rather is linked to modes of expression. I cannot simply use Microsoft PowerPoint for my lecture.

Creative Processes in Evolution

Here is an overview of some creative processes in evolution.

Freezing Time
Limited Spaces
Multicellularity
Sexual Reproduction
Bilaterality
Adaptability

- Learning
- Speed: Optimizing Effortless Processing
- Unpredictability
- Communicability
- Individuality
- Categories ("Whatness")
    - Reduction of Complexity
    - Forgetting (Abstracting)
- External Point of View (Autonoetic Processes)
- Frames (Prejudices)

The first creative process was "freezing time." Deoxyribonucleic acid (DNA) did just that: DNA froze time to conserve situations from earlier to later. In order for DNA to freeze time, "limited spaces" or cells were invented. Plants had cells some 4 billion years ago.

Then, perhaps 700 million to 800 million years ago, "multicellularity" was invented or discovered. This gave birth to the development of brain systems. Next, something happened to speed up evolutionary processes: "sexual reproduction." This opened up new domains.

Subsequently, something quite surprising happened. In all moving systems, "bilaterality" was invented many times independently, possibly under the influence of gravity. This bilateral construction gave rise to 2, 4, or even 8 eyes in some insects. As a result, symmetry was introduced into moving structures, and eventually into the human race.

Then, there was the invention of "adaptability" or "learning." This is a problem for creativity. We are optimized for effortless and speedy information processing. We open our eyes immediately to see something; we effortlessly understand something, effortlessly start a movement. However, this effortless processing is contrary to opening our minds to new ways of processing. These categorical representations somehow limit our ability to see things differently.

In contrast to behavior and predictability, the discovery of excitatory and inhibitory neurotransmitters and their connection in a network — which has divergence and convergence of projections — created "unpredictable" behavior. This in turn has triggered other evolutionary steps.

Thereafter came "communicability," "individuality," and "categories." By categorizing things, we can reduce complexity. The same can be said of forgetting: getting rid of unnecessary information. This is a necessary machinery for constructing categories.

Then, the "external point of view" was discovered in humans. This allows us to have an episodic memory, where we can refer to images from our memory of the past. Lastly, "prejudice" was invented. This is somehow necessary for dealing with nature.

In short, evolution is quite creative.

Some Lessons from Walle Nauta

So, what is the brain machinery? Let me give you a very simple overview of the principle structure of how the brain processes information. Here, I refer to the book *Fundamental Neuroanatomy* by Walle J. H. Nauta and Michael Feirtag.[a] To me, Nauta was the most eminent neuroanatomist of the last century. He spent some time at the Massachusetts Institute of Technology (MIT); and for us young scientists who worked at MIT and who all had a copy of his book, we thought it was the best treaty ever.

[a] Nauta WJH, Feirtag M, *Fundamental Neuroanatomy*, Freeman, New York, 1986.

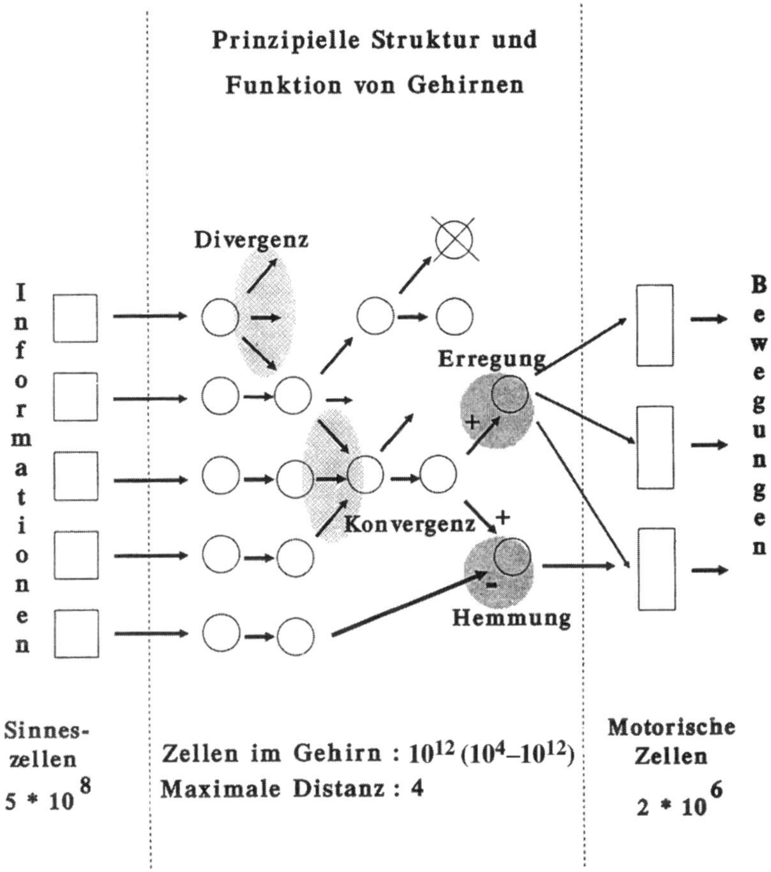

According to Nauta, all nervous systems ever developed, in principle, have the same structure. I believe this imposes constraints on our brain machinery.

There are three types of nerve cells (neurons): sensory neurons, motor neurons, and interneurons. Sensory neurons function as antennae that reach into the outside world, which is shown here on the left column of the diagram, via receptor systems. Visual receptor systems are sensitive to pick up information between 400 to 800 nanometers; and auditory receptor systems, between 20 to 20 000 cycles per second (Hz). Therefore, sensory neurons receive specific information from the physical world and sent it to the central nervous system; however, we are at the same time blind to the rest of the world. Limits are set by our sensory organs.

In my opinion, science is the only way to transcend the limited view we have effortlessly formed about the world. Effortless access to information comes from outside, but the creative process of science has allowed us to transcend or move beyond this effortless view, for example, by inventing telescopes or microscopes.

The second type of brain cells are motor neurons, which represent the output of people's movements. These neurons send information away from the central nervous system to muscles and glands. In humans, there are only about 2 million motor neurons. They are the entire machinery that produce output in the form of talking, moving, etc.

Any nerve cell projects to approximately 10 000 others. The little circles in the diagram represent nerve cells. As shown in the diagram, there is a large divergence as well as convergence of projections. Every nerve cell receives about 10 000 inputs. Now, if each input was independent of each other, then such a nerve cell would have $2^{10\,000}$ potential states ($=10^{3000}$), which is quite a lot.

Because of the architecture of the brain, we have this gigantic distribution of information. By the way, all of these nerve cells are connected to each other through interneurons, the third type of nerve cells. For this purpose, there are excitatory and inhibitory neurotransmitters.

Now, let us ask ourselves this question: what is the maximal distance between any of the nerve cells, of which there are about 500 billion in the brain? The maximal distance is four steps. So, any nerve cell in this network can influence another one by just four intermediates, excitatory or inhibitory. This implies that there cannot be any kind of independence in processing. For example, the information from local excitatory visionary processes reaches the frontal lobe in four steps. This simple fact means that there is no perception that is independent of emotional evaluation or of the relationship to memories.

The problem we run into here, in our philosophical tradition, I think, is Plato's responsibility. We believe that the words we use are self-contained modules; thus, when we use words like "consciousness," "memory," or "perception," we immediately imply that there are independent processes in these modules. Of course, they are modules, but no module can act independently.

So, we now know that there is no activity in the creative process that can be thought of as independent of one's emotions. This is because of the brain structure, and we should not be misled by the words we are using. Therefore, it does not make sense to me to ask such questions as, "Where is the site of consciousness?" It is simply not a relevant question because of the architecture of the brain.

We have established that we enter the world with many synaptic contacts, with potential semantic contents. We enter the world with potential interaction. Synaptic contacts are genetically determined or given; however, they are not determined forever. Every semantic content, almost every synaptic contact has to be reinforced or confirmed by information processing.

So, during the first 10 years of one's life, the entire matrix of the potential nerve system is either confirmed or not. The things that are not confirmed are no longer effortlessly available. Therefore, "imprinting" is a special format of learning: it is elementary for setting up this matrix, which in turn is the matrix for future creativity. The more hubs are created, the more confirmation we have in different domains, and hence the more richness we have in the interconnecting modules.

As an example, I do not know of any person in the world who can speak a second language without an accent if this person has learned only one language for the first 10 years of his or her life. Of course, there are many people who have learned up to four languages and, if they have a broad repertoire of phonetic competence, can easily learn a fifth or sixth language, without having an accent. So, I think perhaps 100 or so phonic sounds are genetically given, but they have to be confirmed during the early phases of life. We as adults have a limited set of phonetic competence. There is no way we can effortlessly, and without an accent, learn a second or third language.

The same goes for the confirmation in the emotional domain. Emotions, which are to an extent genetically given, have to be confirmed at an early stage so that the richness of the matrix is guaranteed.

This is the story of the conceptual brain.

• • •

Words as 'modules'

> Each sensory system is characterized by a number of modules: in this example, only three such modules are shown for each modality. The different modules may be viewed as the structural or algorithmic correlates of particular perceptual qualities. Such qualities in the visual domain are, for example, color, movement, or, perhaps, faces. For taste, these qualities are sweet, sour, salty, or bitter.[b]

• • •

Now, let's look at the activation of various modules. If these modules do not work, all of the systems go down. This is a state, for instance, observed in deep depression — during depression, activation goes down. This means that people cannot remember things, their emotions become flat, and they cannot initiate anything; even their perception may go down. So, if you are in such a state, it is quite clear that mental creativity is somehow defect. This is the simple consequence.

Every cell in the body obeys or is modulated by the circadian clock. We learn fastest, for instance, between 10 AM to 12 noon, but we remember best between 4 PM to 7 PM. So, every physiological function is characterized by the oscillation in 24 hours. In a sense, we are changing our identity all the time — we become close to ourselves every 24 hours.

The 30-Millisecond 'Brain Clock': Minimal Time

In experiments on temporal information processing, one particular number continuously pops up. This number is 30 milliseconds. Let us call this number the "brain clock." I propose that sensory modules are

---

[b] Poeppel E, "Taxonomy of subjective phenomena," *Acta Neurochirurgica* **57**: 123–129, 1993.

organized by this brain clock: the brain clock oscillates approximately every 30 milliseconds and organizes sensory modules. Put simply, the oscillatory processes in the brain help to reduce complexity.

Let me give you some physiological or biophysical examples. First, I will give you two distinct visual stimuli, and you have to tell me which one comes first and which one comes second. In other words, you have to answer the "before–after" relationship. Experimental evidence tells us that two stimuli must be separated by at least 30 milliseconds for you to tell the time order.

Next, let us do the same experiment with auditory stimuli. Interestingly enough, again, the temporal separation must be at least 30 milliseconds.

Thirdly, repeat the experiment with tactile stimuli. You are being touched twice. Again, to be able to tell the "before–after" relation, at least a 30-millisecond temporal separation is necessary.

Three different sensory modules, and only one number: 30 milliseconds. How is this possible? One way is via the following: during the early phases of information processing on the visual system, neuronal oscillations are triggered. These oscillations create "system states." These elementary processing units are timeless, so information that comes within the time window of approximately 30 milliseconds is treated as cotemporal.

So, the idea is that the brain sets up "timeless zones," which I use as the elementary integration units. By "timeless," I mean that the brain steps out of the flow of absolute time, as Isaac Newton writes in *Principia Mathematica*: "[A]bsolute, true, and mathematical time, of itself and from its own nature, flows equably without relation to anything external."[c]

Within these timeless zones, the brain does not (cannot) distinguish the before-after relationship. Since all the information that comes inside this timeless zone is treated as simultaneous, the zone works like an elementary cabinet or file to reduce the complexity of the world around us. In short, the 30-millisecond brain clock is our time organizer.

[c] Newton I, *Principia Mathematica*, 1687.

Column: evidence of a 30-millisecond timeless zone

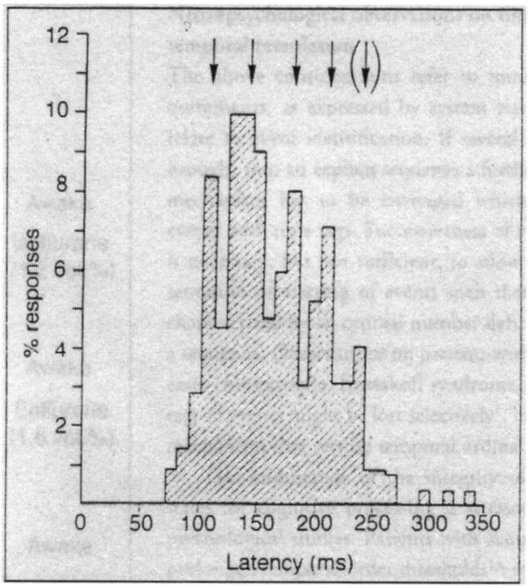

Evidence 1. Histogram on the basis of 463 latencies of pursuit eye movements in three subjects.

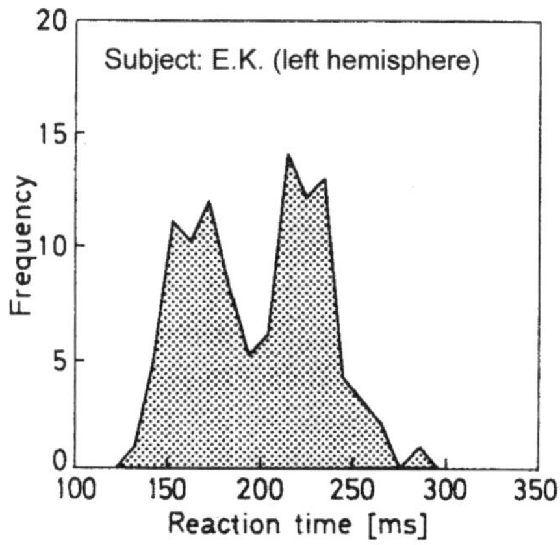

Evidence 2. Histogram of acoustic choice–reaction time for a subject who had to discriminate between optic and acoustic stimuli before the proper response was made.

The two histograms shown above, which are two examples taken from very different domains of research, provide evidence for my hypothesis of the 30-millisecond timeless zone.

Evidence 1 shows the results of an experiment on human eye movements using visual stimuli. Subjects were asked to pursue the movement of a circular target projected on a screen. When the target started to move, a delay or "latency" was observed in the pursuit movement of the subjects' left eye. However, the initiation of the movement was not a continuous process; rather, the initiation process was controlled by the 30-millisecond oscillatory mechanism.

Evidence 2 shows the results of a choice reaction experiment in the auditory domain. Here, the necessary time to distinguish the first–second or before–after relation was also 30–40 milliseconds. However, these choice reaction decisions were not a continuous process; they were discrete, as is observed by the many peaks in the histogram. This "now and then" was, again, a step of 30-40 milliseconds. (Of course, this number is not a physical constant, but may vary; for example, 50-60 milliseconds may have been needed for elderly people.)

Let us consider what happens during anesthesia. The input is still there and information is picked up, so the brain still responds. However, anesthesia takes away the temporal lobe between neighboring units, so system states are not prepared. As a result, no information is distributed in the brain, and there is no way to create conscious events. If you ask the patients what happened during the operation, they would answer "nothing" or "no time." Without temporal organization, without the brain clock, your brain cannot process information.

We have elementary system states with a temporal width of 30 milliseconds.

The 3-Second Window of Consciousness: The Sense of 'Now'

We now come to, what I think is, the most important machinery: the window of conscious activities.

The system states are elementary processing units that follow each other in time. How is their information linked so that we can experience continuity? Usually, we think that the information given by the elementary processing units is related to each other, so continuity is created on the basis of information processing. However, this is wrong, for the brain does it differently.

The brain creates processing units, or "windows of integration," which have a duration of 2 to 3 seconds and are presemantically determined. There is a lot of experimental support for this idea, but only a few examples will be mentioned here.

The first experiment is a very simple one. Let's say a light bulb is switched on in front of you. After a couple of seconds, the light is switched off. The question is this: How long did you feel the light was on? You ask the subjects to reproduce a duration. Now, up to 3 seconds, most people can tell the duration fairly well; however, beyond 3 seconds, people begin to deviate from the right answer. So, it seems that we can somehow keep in mind information up to 3 seconds, but not beyond that.

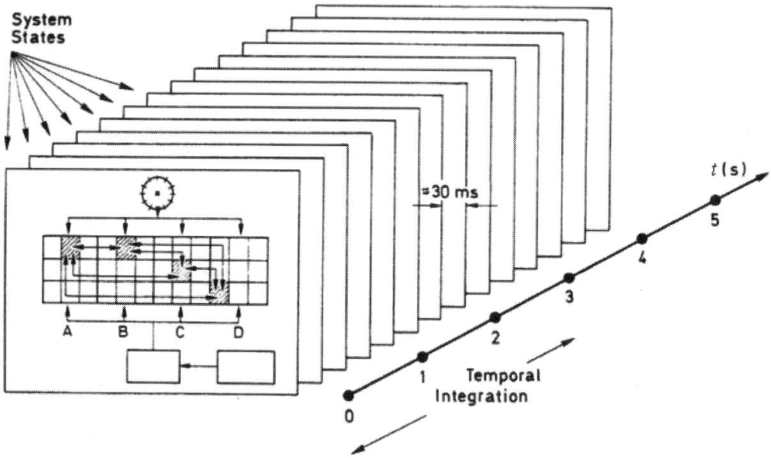

Interestingly, autistic children look directly into the elementary operations of the brain. They reproduce any interval within 3 seconds; sometimes, they go up to 6 seconds. So, they are looking directly into this machinery.

Another example that provides evidence of the 3-second window of consciousness is the time order error. When you compare the intensity of two stimuli (with respect to weight, loudness, brightness, etc.), you have to do it within a window of 3 seconds. If you go beyond 3 seconds, the estimate is always reduced.

Now, let us perform a simple experiment called "cubacu." You process computer-generated syllables "cubacuba…," and ask the subject to try to prevent the change of concepts. You will hear "cuba" for 3 seconds, then suddenly you will hear "bacu." You cannot prevent "bacu," "cuba," "bacu" — it switches all the time.

As a similar example, take the Necker cube, which is shown below. The Necker cube has two perspectives. You cannot prevent seeing either one.

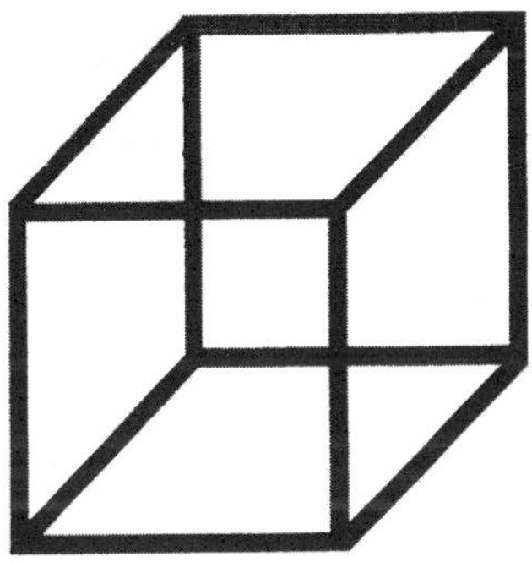

Now, let us conduct a creativity experiment using two Necker cubes, as shown below. You can view this diagram in five different ways. On the left side, you have two altering perspectives; the same goes for the right side. You can also see it as a flat picture. Try to voluntarily control these five different perspectives. It is actually great fun, but requires concentration; and I think this is a wonderful exercise to open your brain for perception.

This is a strange kind of science: collecting information from different domains and trying to integrate it into one picture.

The next example involves the accentuation of a metronome. Play a metronome with tempo = 120. You are given almost identical intensities, but you can easily accentuate them: "ta tang ta tang...," or "ta ta tang ta ta tang...," for instance. Then, slow down your metronome to, say, tempo = 80. Obviously, it becomes harder to maintain your rhythm. Your "ta...ta...tang" is no longer felt as one solid unit. It is hard to bind together a musical note beyond 3 seconds.

The temporal segmentation of speech is yet another example. In all languages of the world, people who talk spontaneously produce 10 to 40 syllables followed by an interval of up to 500 milliseconds, which is often nonvocal or nonlinguistic, thus bridging sequential statements. This is also reflected in poetry: all poetic verses in all languages have such a window. Try reading this aloud: "Shall I compare thee to a summer's day?" This is the first line of a famous Shakespearean sonnet. Look at your watch, and you will see that it takes approximately 3 seconds to read it. Now, read the following:

> Shall I compare thee to a summer's day?
> Thou art more lovely and more temperate.
> Rough winds do shake the darling buds of May,
> And summer's lease hath all too short a date.[d]

Each spoken line lasts approximately 3 seconds.

Next, the duration of music motif is used as an example. If you listen to Beethoven's Fifth Symphony, the famous motif lasts about 3 seconds. Likewise, if you go to a Noh play in Tokyo, the drama is segmented with a 3-second time structure.

All of this implies that there is a common window of 3 seconds, which is presemantically given as a temporal stage whereby vision, audition, memory, motor planning, and speech take place. I take this as the pragmatic definition of the state of being conscious.

[d] William Shakespeare, "Sonnet 18."

The most important author on time is still, of course, Augustine. In his 11th book of *Confessions*, which was supposedly written in 397 AD, he says the following: "What is time? If no one asks me, I know the answer; if I want to explain it to someone who has asked, I do not know." [e]

We cannot talk about the future or the past; the only thing we have is the present. Everybody knows the present, but it is nothing like the past or the future. We should not say that the past, the present, and the future are three different times; instead, we should say that there are three types of the present. In a sense, the presence of the past is "memory," the presence of the present is "experience," and the presence of the future is "expectation." This is a pragmatic, experimental definition of the Augustinian present, which is approximately 3 seconds. These 3 seconds are the stage for human consciousness.

So, 3 seconds is the window or logistic frame within which everything happens. Moreover, this time frame is prefixed or presemantic.

The continuity of thought, on the other hand, comes about by a semantic connection. The feeling of continuity is a secondary consequence of the semantic process. In other words, the so-called "flow of time" is created by (effortlessly) connecting successive 3-second "nows" by their semantic contents or meanings. If you are under the influence of alcohol or suffering from schizophrenia, you may actually cut this semantic continuity, and so you suddenly may not know what is going on.

The fact that this flow of time happens effortlessly indicates that the brain has created the machinery to create continuity on the bases of evolutionary processes. Therefore, the 3-second window is necessary to be logistically used as a stage in order to represent information that is semantically connected (at the same time, however, the brain machinery is wired to look backwards, to retrospectively reflect on what has happened).

Why do we only have one "now" at a time? Why don't we have many "nows"? Well, my speech is segmented to 3 seconds. I do not have to finish a sentence because you can anticipate what I'm going to say, can't you? So, it's highly economic to create only one present. Evolutionarily speaking, it is better to have only one present.

---

[e] Augustine, *Confessions*.

Complementarity and Creativity Revisited

When we talk about creativity, we are actually referring to the brain machinery. To make sense of this machinery, I will make use of a new concept I acquired from Dr Rhunau, a theoretical physicist who defined complementarity as a descriptive principle in quantum mechanics. However, what I suggest here is complementarity as a generative principle in setting up thoughts, as shown in the following list. The list could be much longer, but only a few complementarities will be stated here.

Complementarity as a Generative Principle

Sensory Data (Chaos)	Elementary Units (Complexity Reduction)
Local Features	Global Invariants
Bottom-up (Details)	Top-down (Schemata)
Temporal Frame (Presemantic Integration)	Content ("What it is")
Perceptual Identity (Stability)	Perceptual Change (Dynamics)
Perceptual (Effortless) (Present)	Conceptual (Time-Consuming) (Retrospective, Anticipative)
Chance	Meaning
Segmentation (Modules)	Integration (Coherence)
Implicit	Explicit
Verbal	Pictorial
Efference Copy	Reafference
Answer	Question

We have a tendency to explain things monocausally, but I think we should look at things from a complementary point of view.

The first item on the list is incoming sensory data (which I talked about earlier) that are distributed to the brain. We need this data, but we also need the elementary processing units to reduce the complexity that comes with the data.

Next, we have the bottom-up top-down machinery. The best example I can think of is prosopagnosia. Prosopagnosia refers to the incapacity to distinguish faces. Patients with surface injuries in the right hemisphere of the brain can no longer recognize faces, including their own. When they look at a face, the schema of the face is still there, but the raw information or details needed to identify the specific face and distinguish it from other faces is no longer possible. This is another complementarity: you have the schema, but then you also need the local information or features to identify the specific object.

Another complementarity is the previously-discussed temporal frame (presemantic integration) and content. So is perceptual identity and perceptual change — this is perhaps most important for my present reasoning. The 3-second window is a way to maintain the identity of a perception or thought. The identity of perception is maintained using constancy, such as color constancy, brightness constancy, etc. However, you do not want to be frozen with this identity; you want it to change. So, after 3 seconds, the brain opens up for a new perception, a new identity. In this way, you have the complementarity between the maintenance of identity and dynamic change.

We also have perceptual and conceptual complementarity, which has already been explained. I started my talk with the complementarity of chance and meaning. Our discussion here is based on the complementarity of answer and question. Another example is efference copy and reafference. If you plan something, you want to solve the problem. We always have a complementarity between the plan or efference copy (what to do) and the reafference (information coming in). Thus, we have a self-monitoring system on the basis of complementarity.

I believe that complementarity should be the way to think about our brain, our mind. So, finally, we have to be serious about creativity: we have to develop a taxonomy of creativity. What could be potential domains of variables that have to be recognized if you want to have an education program to improve one's creativity?

Creativity is used to create knowledge, and we have different knowledge systems. There is explicit knowledge; but there is also implicit knowledge, such as the development of heuristics. In addition, there is visual knowledge (which consists of concept, perception, and topological modes of knowledge), episodic memory knowledge, etc. These various knowledge systems have to be included in such a taxonomy.

We also have to consider physiological constraints to creativity. For example, I have to walk around; otherwise, I cannot think. Actually, I cannot talk if I sit; I have to at least stand.

The physical layout of MIT has a maximum radius of 50 meters in order for people to meet each other and form creative groups. The internal hall at MIT is constructed in compliance with this guideline, so that someone can always meet someone else and start a discussion.

Very simple things like office structure also affect creativity. I once gave a talk at a company on how to arrange chairs and tables for the office. How should they be arranged? Where should the doors and windows be? I think these physical constraints are completely underestimated.

What we are discussing here are the expressive modes of creativity. Creativity is not an abstract activity; it is related to talking. The German poet Heine wrote about the gradual development of thinking while talking. According to him, if you have thought about something but cannot come to a conclusion, grab anybody and talk to the person; somehow, the mode of talking about it will solve the problem. The same goes for other modes of expression such as writing, thinking, playing, building, etc.

Social factors and time pressure also affect creativity. You might want to impress somebody, so that might trigger creativity. Of course, people have different evaluations of their own creativity (for instance, some men who want to impress ladies think that they are brilliant, when they are actually stupid).

Furthermore, there are organization aspects to creativity. For instance, you may belong to a group, but at the same time you may be autonomous. This is, again, one of the complementarities that characterize creativity.

Finally, if you want to develop a taxonomy of creativity, which I strongly recommend, you have to think of the different fields. I think that the creative act may be different for artists and poets, scientists and politicians, and educators and mathematicians.

Well, this is all I have to say. Thank you very much.

Further Reading

Poeppel E, *Mindworks: Time and Conscious Experience*, Artin T (trans.), Harcourt Brace Jovanovich, 1988.

In Search of Achilles' Heel

My favorite aunt died of cancer. Cancer seems to be one of the most difficult diseases to be cured.

Once every week, I used to drive my aunt to the hospital for her chemotherapy treatment. At first, her enemy retreated. The overall mass of the cancer was dramatically reduced. However, the doctor was not happy. He knew what was going to happen next.

The cancer kept coming back; it was robust. The chemotherapy did not seem to work any more. The doctor told me he was sorry. Finally, one day, my aunt succumbed to the evil.

We lost the battle against cancer. We continue to lose the battle today. However, Dr Hiroaki Kitano thinks that he can change the whole situation. He thinks we can fight back. He proposes that an effective cancer therapy comes from a theoretical consideration in the field of systems biology:

> To understand biology at the system level, we must examine the structure and dynamics of cellular and organismal function, rather than the characteristics of isolated parts of a cell or organism. Properties of systems, such as robustness, emerge as central issues, and understanding these properties may have an impact on the future of medicine.[a]

---

[a] Kitano H, "Systems biology: a brief overview," *Science* **295**: 1662–1664, 2002.

First proposed about half a century ago by Norbert Weiner, the systems approach to biology is coming back. Recent advances in molecular biology have made it possible to apply Weiner's ideas to the real world of living things.

What is "robustness"? For example, Achilles, the mighty hero of the Trojan War, was robust. He possessed many fighting skills, and could kill his opponent in many ways. In systems biology, this skill is called "heterogeneity." Achilles had heterogeneous fighting skills. Also, when Achilles had to fight a strong enemy, he might retreat for a while; but since Achilles was such a good soldier, he could learn during the battle, change or slightly modify his tactics, and finally overwhelm the enemy. In systems biology, this is called "feedback."

So, a robust person or system seems to have heterogeneity and feedback abilities.

However, there is a catch. Achilles died because Trojans shot him in the heel. That was the only fragile part of Achilles' body.

This ancient tale teaches us that nobody is perfect. In systems biology terms, there is a tradeoff between robustness and fragility.

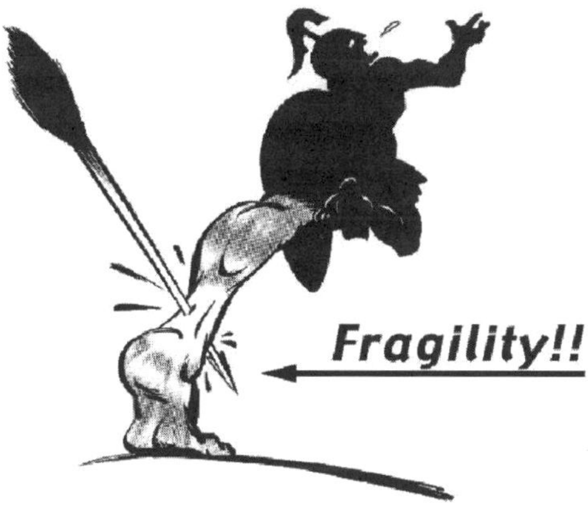

So, the right strategy to fight cancer seems to be to look for the Achilles' heel of cancer. In fact, Dr Kitano believes that as a biological system (such as cancer) becomes robust, a fragile point always appears: "Viewing cancer as a robust system with potential points of fragility opens up new strategies for the development of drugs and therapies."[b]

Interestingly, any robust system has the same bowtie-looking structure. This seems to be a universal feature of the robust system in general.

The entire discipline of systems biology is still in an embryonic stage. However, Dr Kitano hopes to one day find the Achilles' heel of cancer.

---

[b] Kitano H, "Tumour tactics," *Nature* **426**: 125, 2003.

Robustness in Biological Sciences

Hiroaki Kitano is the director of Sony Computer Science Laboratories, Inc., Japan, and the director of The Systems Biology Institute, Japan. He is also the project leader of the ERATO-SORST Kitano Symbiotic Systems Project at the Japan Science and Technology Agency, which has an international research team based in Tokyo and at the California Institute of Technology in the US.

He received his PhD in computer science from Kyoto University. He has also been a visiting researcher at Carnegie Mellon University, USA. He is a strong proponent of systems biology, and is interested in understanding the robustness of biological systems and its medical applications, particularly in cancer. His team has also played an important role in creating software platforms for systems biology, such as the systems biology markup language (SBML) and the systems biology workbench (SBW).

I am working on molecular biology, peculiar tumor biology, and diabetes. One of the things I am interested in is how complex systems like cells and our bodies account for high functional robustness of diseases such as cancer and diabetes.

Let me start with a very simplified example: mycoplasma are very small microorganisms that are only about four hundredth of the size of genes. They are supposed to be very fragile against the changing environment, but were recently found to be much more robust than we thought. As another example, equalized bacteria are about four thousandth of the size of genes, and are able to adapt to a very broad range of environments. It is the equalized evolvable complex that increases robustness.

This same concept applies to the gene system.

Boeing 777 Versus Wright Brothers

Let us compare the Wright brothers' aeroplane with the Boeing 777. The former carried only 200 to 300 components, but the important thing was that it could fly. The Boeing 777 has two million or three million components, and its basic function is also to fly in the air.

What do you think is the difference between the Wright brothers' aeroplane and the Boeing 777? Some people may say that their size is the major difference; but as long as you have a big size, you can have big components. Therefore, the size does not really matter. The big difference is actually their robustness: their stability during the flight. The Wright brothers' aeroplane was able to fly, but it was a very unstable system. Most commercial modern airliners, on the other hand, are extremely stable. You are safe to fly (unless the aeroplane is hijacked by terrorists).

So, the big difference lies in those mechanisms that enable robust flight. A big difference, indeed.

One of these mechanisms is the feedback control. To be sure, there are external and internal perturbations during the flight; however, there is an automatic flying control system (AFCS), which adjusts the internal mechanical and airship noise and the aero-convection. AFCS maintains the flight path by means of negative feedback mechanisms — hence, the feedback control. Its flying mechanisms maintain the robust flight path management. We have automatic flight control computers on board.

Now, there is a redundancy involved here. Most commercial aeroplanes carry three computers, in case one of them fails. But, we still need a specific condition to increase robustness: all three computers have to be implemented differently. We need identical functions, but different implementations. This is the key to robustness.

Even if you carry three computers, if all three have the same hardware and the same software, then they will probably all go wrong at the same time. In order to avoid such a situation, we need "heterogeneous" implementation. Backup computers must do the same job, but in a different way.

This kind of robustness is widely observed in nature — even at the molecular, cellular, or tissue level. The two primary manifestations of robustness are adaptation or insensitivity to inparametric and parametric variations. For example, adaptation in bacterial chemotaxis means that when the bacteria swim to the nutrients, they can sense differences of 10^6 (which is a very large dynamic range), yet they can accurately swim to the nutrients and avoid the chemo-repairs. So, for them, the intensity of the nutrient or chemo-attracter does not really matter.

We know that bacterial chemotaxis is robust. Even if you change the expression level of the gene, the behavior of the bacteria stays intact (although the adaptational type will be different). So, the bacteria exhibit strong robustness.

How is this achieved? With the systems control, feedback control, feed-forward control, heterogeneous redundancy, and — of course — modular design and structural stability. These are the mechanisms behind the manifestation of robustness, and are observable in a cellular system or in tissues.

One of the physics-based observations is Jim Carson and John Doyle's highly optimized tolerance (HOT) theory. They argue that robust complex systems show robustness against common perturbations, but reveal extreme fragility when faced with unusual perturbations. In other words, to be robust against general perturbations, the conventional division of economy must be extremely fragile against unexpected perturbations. So, the total amount of robustness is still debatable. If you want to be robust against one aspect, you are exposed to fragility in another aspect.

For example, the Boeing 777 is very robust against atmospheric perturbation, and is considered to be much more robust than the Wright brothers' aeroplane. But, these modern jet airliners are extremely fragile against very unusual perturbations, e.g. a total power failure. When the power is lost, there is no way we can fly this aeroplane. In contrast, the Wright brothers' aeroplane, although very fragile against conventional atmospheric perturbations, could fly whether the power was on or off because it did not use any electrical controls.

So, the Boeing 777 enhances aerodynamic robustness via the modern, sophisticated control technique; however, because of this, the system is extremely fragile when the power supply is gone. Obviously, there is a tradeoff here.

There are several issues regarding the tradeoff of robustness and resources, or robustness and fragility. If you want to increase robustness, you need a redundant system; for example, you need two or three (maybe more)

computers. You need resources, but this might pose a problem. To enhance robustness, you need more resources or more ecological need, which is limited in most cases.

Let me give you another example. The F117 A Nighthawk is a fully computer-controlled stealth aircraft. Its flying dynamics is so unstable that humans cannot even fly the aeroplane manually, and yet the reason this aircraft is so successful comes precisely from the complete machine/computer control. Because of the instability of flight dynamics, it can actually maneuver much quicker than any conventional aircraft. The F117 can be tuned much faster than any other aircraft because it is unstable. Therefore, there is a tradeoff between the stable flight path and instability. If you want to have a more stable flight, then you lose the maneuverability.

The same thing happens in an automobile; for example, in Formula One racing. If you want to have a car which is more stable in a straight line, then you lose maneuverability at the corners. Conversely, if you would like to optimize the design of the car to maneuver in a high-speed corner, then you lose the stability, thus making it very difficult to keep the car straight at high speed. To make a quick turn, you introduce instability. This is another example of a tradeoff.

Cancer Therapy and the Origin of Robustness

I am currently working on a tumor system. It is very difficult to cure cancer. As doctors, we prescribe a regimen of chemotherapy, radiation therapy, and whatever else is likely to remove the cancer cells. Sometimes, if we are lucky, we can cure patients, but usually it is very difficult. We may observe tumor mass reduction after the first round of therapy, but it will then rebound, and eventually metastasis occurs.

One of the hypotheses I devised is based on robustness. If you look very carefully, there are two types of robustness of tumor cells: functional redundancy and feedback control. Functional redundancy is enabled by cellular heterogeneity. Feedback control is observed at the intercellular level and in the tumor–host environment. There are a lot of feedback loops, so it is very difficult to cure cancer.

A conventional successful measure of tumor therapy is the reduction in tumor mass. Doctors try to reduce the amount of the tumor cells. However, my advice is this: "Hey, doctors, that's not a good idea." The guiding principle should be to control the robustness of the tumor, not the size of the tumor, as robustness is a manifestation of abnormality or disorder in our body.

When a tumor grows, it creates a mass. Inside the tumor, it is very difficult to get oxygen, thus causing hypoxia. However, tumor cells can still metabolize by using glycolysis rather than the oxygen-dependent tricarboxylic acid (TCA) cycle to produce energy or adenosine triphosphate (ATP). This, by the way, is a normal cellular response. Normal cells obtain energy by switching the strategy, which is essential to our survival. Moreover, it is not only our cells that are doing this, but all sorts of organisms. This is a very fundamental robustness: one function of ATP production has two recommendations, just like an AFCS in an aircraft.

The same thing happens in diabetes. Diabetes is actually a part of normal bodily response. It is merely a feedback against environmental changes, but the robustness is working against us.

Cancer tactics

When the degree of heterogeneity is low (such as in early-stage, nonsolid tumors), it would be best to avoid broad-spectrum cytotoxic drugs that may increase heterogeneity, and instead look for a drug that targets a specific molecule.

When the level of heterogeneity is high, there is a choice of two strategies. The first is to find a method that actively reduces the diversity of mutations among tumor cells. The alternative would be to guide the malignant cells into dormancy and slow regression.

Now, the question is where this robustness comes from. Is it an inevitable consequence of any complex biological system? Here, we need the concept of "evolvability," which was conceived by Gerhart and Kirschner (1997).[c] Evolvability, in a word, is the capacity to produce diverse phenotypic variations. Gerhart and Kirschner have presented several possible mechanisms behind evolvability.

---

[c] Gerhart J, Kirschner M, *Cells, Embryos, and Evolution*, Blackwell, Malden, MA, 1997.

One mechanism is "weak linkage." Processes are weakly rather than tightly coupled, so a change in one part or component will not seriously affect the other parts. As a result, the system can cope well with changing environments.

Another mechanism is "compartmentalization." Whatever is happening in one region in a static or temporal space does not affect other regions. Therefore, you can exchange one compartment without affecting what is going on in other compartments. This is called "modularity."

Gerhart and Kirschner also argue that the emerging architecture is composed of highly conserved core processes that are coselected with various other processes. The resulting robust structure has the topology of a bowtie.

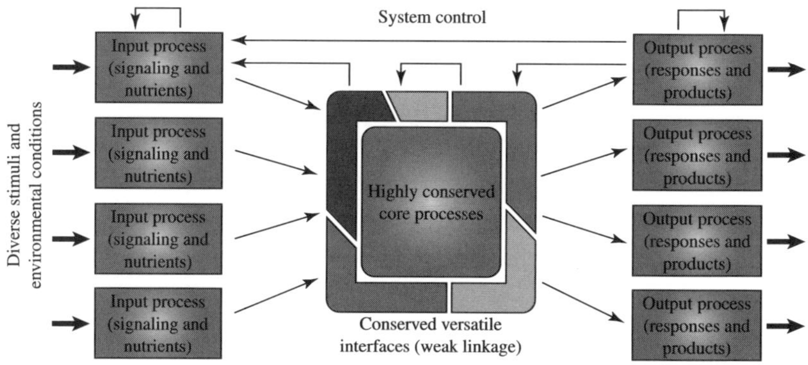

Tradeoffs: A Possible Connection with the Mindset?

Now, I begin to delve into topics related to this workshop. Please note that these are somewhat speculative ideas, not established facts. Nevertheless, I still believe that the concepts of robustness and tradeoffs are useful.

After birth, cells in the brain system demonstrate a high potency: they can turn into any kind of cell. Development is the process of differential inhibition of potency. As the brain develops, the cells lose power to turn into something else. Their functions become fixed. This is a tradeoff between remarkable potency and excessive functions.

The brain is a modular, weakly linked system. It shows an extremely high level of genetic flexibility, so it obviously fulfills the evolvability criteria: core process and diversity. There is a clear tradeoff here between robustness and fragility.

Now, I come to the problem of consciousness. Is qualia a robust invariant of perception? I am interested in early-stage mindset formation. I think the mindset, in a way, produces an invariant; and if the invariant is unstable, then it will be very difficult for things to be robust. Therefore, in a developmental or generation-to-generation scale, evolvability requires very fixed and steady invariance in the core process. At the same time, though, the core process is very fragile or unstable, and thus continues to move dynamically to be other parts of explored things.

The mindset may be the formation of a robust invariant. It could be the building block of complex cognition. If the mindset is not formed, maybe we would not be able to create complex cognition.

Creativity is probably an expression of this concept. In order to have creativity, we might need a core process that is rather fixed. Here, we have a knowledge tradeoff because the formation of a mindset impairs creativity.

You have to fix or stabilize the core process to explore new things. Then, if the same thing is required in the high-cognition level, you might need to have a fairly conserved, stable, set invariance in order to explore things. Again, there is a tradeoff here.

• • •

Discussion

Snyder: I was fascinated with the idea of a tradeoff. Is there any mathematical explanation or discussion? I don't know how to mathematically quantify this concept of the mindset. But, the tradeoff between the penalty you paid, or having the mindset, and the gain must somehow be a way of mathematics.

Kitano: Rightly so, in a narrow context of robustness, in a modern control theory; that is, the concept of controllability and stability analysis. You have all the attenuated dimensions of space and how much you can push.

That actually captures only the stability of negative feedback control. So, if you have a redundancy, or a performance of robust tradeoffs, that is out of scope. If you have negative feedback, you resist the gain, so you sacrifice performance in one sense. But, that part is not yet quantified, so we need some new mathematics to really have a general idea of the robustness tradeoffs. That is actually what I am working on right now.

Snyder: How do you do that in the cancer problem?

Kitano: You have genetic diversity in a tumor cell. So, one of the things that we are actually doing right now, with a couple of hospitals, is to actually measure heterogeneity. We can quantify this because we can quantify diversity at the genetic level.

Snyder: OK.

Kitano: Then, as the tumor progresses, how does this measure increase or decrease? Or, if you have a high level of heterogeneity, is the tumor going to be more robust against chemotherapy? The preliminary evidence suggests that my theory about heterogeneity may be right in myeloma, which is a kind of leukemia. We tested two types of myeloma patients: one type had been treated with chemotherapy for more than 5 years; the other, less than 3 years. The latter type of patients had high levels of heterogeneity, which increased as chemotherapy went on. However, for the other type of patients, heterogeneity did not increase, despite the chemotherapy treatment. So, we can actually quantify heterogeneity.

Snyder: What about the brain? The brain is another problem. Do you think that the brain is not just a mathematical tradeoff phenomenon? Why doesn't the brain adapt the mindset in parallel with flexibility? I don't understand why the brain doesn't do that. There is almost some kind of economy going on. It is not just a tradeoff.

Kitano: It is not only heterogeneous things that increase robustness; you need resources as well. For example, if you have links in a tile, performance may be sacrificed because you use less neurons or because you need glucose.

• • •

The level of cognition is very difficult to quantify. It is easier at the cellular level because you have a system. The cell cycle consists of two major cycles, e.g. two protein levels, and you can actually change one protein level. This cell cycle goes on, so you can see that it is robust; and you can measure it, and see that it is also very fragile. So, you can actually compare and quantify the robustness and fragility. We can do this because we have good measures like protein activation or protein concentration. Can we do this at the cognition level, too? That is the big question.

Further Reading

Kitano H, "Tumour tactics," *Nature* **426**: 125, 2003.

Kitano H, "Systems biology: a brief overview," *Science* **295**: 1662–1664, 2002.

Kitano H, "Biological robustness," *Nature Reviews Genetics* **5**: 826–837, 2004.

Do Infants Dream of Baby Sheep?

What is the original meaning of the word "infant"? It comes from the Latin infans, *which literally means "unable to speak." So, infants by definition do not speak. Usually, infants begin to utter words at around 12 months of age. They begin to toddle at around the same time. Every child experiences several stages of development: a fetus, an infant, a toddler, and finally a child.*

The big question is this: how do we know the psychology of infants? In other words, do they count baby sheep when they cannot sleep? Given that infants do not speak, it is of no use asking them verbal questions. Is there any way of knowing how infants feel, what they think, or the reasons behind their behavior?

Professor Philippe Rochat does just that. He is an expert on the developmental psychology of infants. A gentle man from Switzerland, Professor Rochat has tender eyes and speaks with a mild French accent. His answer to the big question is called the "habituation/dishabituation method." Professor Rochat describes the paradigm as follows:

> Suppose you are interested in knowing whether infants perceive colors, and in particular whether they can differentiate between primary colors such as yellow and red. You can figure that out by presenting infants with one of the two color cards over and over again. You will time infants' gazing at the card, and when you notice that they pay no more attention, you will flash the other color card. If the infants recover their visual attention, this suggests that they discriminate between the two colors.[a]

[a] Rochat P, *The Infant's World*, Harvard University Press, Cambridge, MA, 2001.

So, developmental psychologists do have means to infer the inner world of infants.

Professor Rochat says that, in order to understand the infant's world, we need to consider three aspects: the self, the object, and other people.

Infants first try to explore their own bodies. They discover the so-called "ecological self." The ecological self is the sense of their own bodies interacting with the environment. The ecological (physical) self should be distinguished from the interpersonal (social) self. This is close to the two basic kinds of self proposed by William James: "I" and "me."

Infants first explore "I," and then move on to discover "me." Infants then move on to explore physical objects around them.

Finally, infants begin to recognize other people. They gradually develop the sense of other people's minds, i.e. how they feel and why they behave the way they do. Infants recognize that the people around them are not merely physical objects, but have feelings and intentions as well. Infants learn to read people's minds — hence, the beginning of the reciprocal or dyad relationship: "you" and "me." This crucial step comes at around 2 months of age. Professor Rochat calls this stage the "revolution

What happens at 9 months?

with a smile" because it is characterized by the emergence of the first socially elicited smile. Parents recognize a "person" in their babies.

Of course, the three key elements — the self, the object, and the people — are interrelated: "Prior to this transition, infants cannot be taught because they do not yet understand that others are trying to teach them."[b] *Therefore, we need to consider the triad exchanges. At around 9 months of age, infants begin to understand the relationship between themselves, objects, and other people. Professor Rochat calls this stage "the nine months revolution." This new stage of development may be called the "revolution with a blanket." Why? Professor Rochat will explain to you in the following lecture.*

[b] Rochat P, *The Infant's World*, Harvard University Press, Cambridge, MA, 2001.

Creativity in the Infant's World

Philippe Rochat is a professor of psychology at Emory University, USA.

First of all, I want to thank you for inviting me. I hope my talk will help you. It is a great pleasure, and I really enjoy this workshop.

I would like to propose that a major bottleneck to creativity is "self-consciousness." This is a take-home message. What do I mean by self-consciousness? It is the understanding or awareness of being evaluated and perceived by others.

New Guinea, 1975

To give you a sense of the kind of research that we are doing and the origin of self-consciousness, I am going to start with an observation reported in the early 1970s by a visual anthropologist named Edmund Carpenter. Together with his team, he visited natives of New Guinea who lived in very dense forests with no mirrors, no reflecting devices, and murky water all around; basically, having the experience of themselves through shadows and things that they were able to directly modify.

Edmund Carpenter and his team introduced video cameras, Polaroid cameras, and mirrors; and systematically recorded the reaction of native adults confronted, for the first time, with a clear view of themselves. What he observed was a systematic expression of great anxiety. After the first frightening reaction, the adults became paralyzed, covering their mouths and hiding their heads. They stood transfixed, looking at their own images, with only their stomach muscles betraying great tension. So, there was a negative or adverse reaction to the encountering with the self.

You might ask why. Why did these people behave like that? You might have anticipated that they would have been delighted, but this was not the case. I think this reveals a fundamental aspect of the human condition. Do you have any idea why? What came to their minds?

• • •

Steels: What comes to my mind is the difference between what they thought they looked like and what they actually saw in the mirror.

• • •

That is what I think, too. Not only did they recognize themselves in the mirror, but they also understood how other people saw or perceived them.

My mother is 84 years old. I just visited her. She is all broken down and has all the diseases you can imagine; but if you look at her eyes, she looks like she is 15 years old, and she feels like she is 15 years old. This extraordinary discrepancy — how one feels inside and what one projects to the outside world — that we all have is, I believe, a fundamental aspect of the human condition.

Let me quote from a book I found that captures very astutely the emergence of self-consciousness:

> If no one is watching them, nothing is really happening to children. This is not some philosophical colloquium like that one of the tree falling in the forest and no one is hearing it. If you were very small, you actually understand that there's no point in jumping into a swimming pool unless they see you doing it. The child

> crying, "watch me, watch me!", is not begging for attention; he's pleading for existence itself.

This quote captures very nicely what I am talking about. Without an audience, we are reduced to nothing. When we write a book, paint, or play music, we have an audience. We are fundamentally and basically social. We cannot think about commission outside of others. Similarly, creativity cannot be conceived as an isolate, solitary, autistic kind of activity. There are always people out there.

Would You Like a Rouge Spot on the Forehead?

The question that we have been addressing for some time now in the lab is how to go from self-perception — perceiving the body by breathing, crying, or moving — to recognizing oneself in a mirror, which happens at around 14 months of age.

Is everyone aware of the rouge task? Usually, we tap the forehead of a child and leave a big sticker under the hairline, so he or she does not feel it. Then, we confront the child with the image in a mirror; the test is whether the child tries to move or touch the sticker. This test was invented by Gallup *et al.* (1971)[c] to see if chimpanzees could recognize themselves in the mirror.

[c] Gallup GG Jr, McClure MK, Hill SD, Bundy RA, "Capacity for self-recognition in differentially reared chimpanzees," *Psychol Rec* **21**: 69–74, 1971.

It is observed that, at about 14 months, children do touch or try to remove the sticker. This seems to be quite universal across cultures. More interestingly, mere self-recognition develops very rapidly into a manifestation of embarrassment: at about 18 months, children try to not only touch or remove the mark on the forehead, but to also vanish from the public eye by hiding their faces and looking downwards. We tested this experiment on children up to 3 years old, and the reactions were similar. Children recognize that they are different from others, and do not like it. There is this propensity of the self, which is related to Allan Snyder's talk about conformity.

We also tested the children in our lab in front of the mirror when everybody in the lab had a sticker on the forehead. In general, there is a lack of embarrassment when the sticker is created to appear normal in the lab. The children do not get embarrassed if everybody else has a sticker on the forehead, and seem pretty happy to be the same as other people.

Furthermore, some children in our experiment clearly experience the equivalence between the other and themselves. So, upon discovering this, instead of taking away their own sticker, the children take the experimenter's sticker and put it on their own foreheads.

• • •

Snyder: Is that before they look in the mirror?

Rochat: No, the mirror is lifted. So, it is really the experience of the normal: we all have stickers, and we are all part of the same clan. This fits perfectly with the idea of affiliation, the idea that we are not different.

Snyder: It's amazing.

• • •

In another experiment, we tested two children: a 2½-year-old girl and her 4-year-old brother. We asked her brother to sit behind her and to keep silent, while we tested the girl first with a sticker. They both did not know that they had a sticker until they saw it in the mirror. However, when the boy looked at the mirror and realized that he had one, he shook his head — this is the crucial point — he wanted to take it off, but then he decided not to. He was very much aware or self-conscious that he had a sticker on, and his first gesture was to try to remove it; but then, he very quickly decided not to, since everybody else had one as well. So, look at the little boy and the inhibition: he realized that he also had a sticker, he shook his head, and then just let it go. I think this is pretty interesting.

• • •

Steels: He is bright.

Snyder: And he fits so well.

• • •

Now, creativity is based on the fact that we are different. Perhaps creativity is trying to break this need for others, this need to align with others. So, in order to be creative, if I follow Allan's ideas, uniqueness must be secured. Think about your own family, Allan! All three brothers have won big prizes and are famous — this is, I think, a fact of the matter you had highly driven, passionate, eccentric parents as role models. You wanted to be accepted by your parents by being out of the ordinary. If you were ordinary, you might have been rejected by your parents!

• • •

Snyder: I think all these ideas about individuality and the need for others, etc., are brilliant.

• • •

We Are Made to Differentiate from the Start

For a long time, pioneers of developmental psychology like Piaget and William James have had this idea that babies are born in a state of confusion or duality in the world, and that they progressively distance themselves from the world. In other words, we gradually objectify the world and differentiate ourselves as independent entities from things that happen outside to eventually become philosophers and scientists, among others, as adults.

However, my research — and infancy research in general — shows that this idea is wrong. We start with the means to differentiate ourselves from the world to begin with. Babies, from birth onwards, experience their own bodies as differentiated entities in the world. I will briefly show you how.

The first object of investigation and exploration is the physical body. Babies move, explore, and touch their own bodies; they hear themselves. This is a primary experience which occurs very quickly by serendipity or fortunate accidents, as described by Piaget (1953).[d] Babies have an affect or influence on things, and they discover their own affectivities in the world. They are agents in the world.

For example, when my daughter was 3 months old, I attached a mobile above her crib one day while she was laying in it. This is what I caught her doing: she lifted her left leg up and discovered the mobile, and then systematically moved the mobile to create an interesting visual

---

[d] Piaget J, *The Origins of Intelligence in Children*, Routledge & Kegan Paul, London, 1953.

experience. What is interesting here is that she kicked, stopped for a while to watch, and then kicked again. She kept doing this for 45 minutes. She was completely mesmerized, not by the mobile, but probably by her affectivity on things in the world.

Before 3 months, babies make perceptual experiences that are unique, thus specifying the body as being differentiated, situated, and agent. I will now demonstrate this to you.

The first thing that is checked at birth is that the baby's lungs are working; if the baby is not crying, we typically hit the child so that he or she will start to breathe. The sound production is primary, and so only the baby can make the experience of himself or herself crying, which sounds very different because of the mixed auditory feedback with proprioception (your "body image"; i.e. the system by which you know where all of your limbs are in relation to the rest of your body, and which informs you of your own bodily movements) and the very complex backed-up stimulation.

As another example, when kids bring their foot in contact with their hand, they make the experience of the so-called "double touch." The foot feels the hand and, simultaneously, the hand feels the foot. This is a perceptual, unique experience, which specifies the body as a differentiated entity in the world.

Do babies differentiate between the self-touch, double touch, and single touch? To determine this, we conducted a very simple experiment based on the fact that, if we stimulate a body region of newborn, typical, healthy babies, they will ordinarily try to suck in the direction of the stimulation. This is a very robotic response that we can expect from every newborn. We tested newborns that were less than 24 hours old, comparing the routine responses in the case of self-stimulation (when they brought their own hands spontaneously to their face) and when the experimenter stimulated a region of their face. The question was whether there was a differential moving response between self-stimulation and other stimulations (i.e. double touch or allowed stimulation). We found that babies groped significantly more to the stimulation by the experimenter's finger compared with self-stimulation. This very simple observation proves that babies do make perceptual discriminations between self-stimulation and other kinds of stimulation.

With respect to causality and affectivity in the world, I would just like to show you one little experiment that, I think, drives at my point: very early on, babies do have a sense of themselves as agents in the world, and it does not seem to be an instrumental conditioning.

This experiment involved 2-month-old babies, each of whom held a pacifier that was connected to a computer, which recorded the positive pressure that the baby was applying on the pacifier. When the child applied pressure above the threshold, the computer generated a sound. Basically, we introduced a musical instrument into the baby's mouth.

For some babies, each time they applied pressure above the threshold, they heard a sound that varied proportionally in pitch. So, it could go like this: tu-tu-tu-tu-tu-tu-tu-tu-tu-tu. This is an auditory analog of the activities on the pacifier.

In another condition, each time the baby pressed the pacifier above the threshold, he or she heard a sound that was not analog to the pressure. So, it could go like this: tu-tu-tu-tu tu-tu-tu-tu. This is a random pitch. In other words, frequency was changed.

We asked two questions: whether 2-month-old infants noticed that the sound was produced, and whether they could discriminate between analog sounds and nonanalog sounds. If they could discriminate, it would mean that babies are engaged in exploration, and that it is simply an instrumental (operant) conditioning.

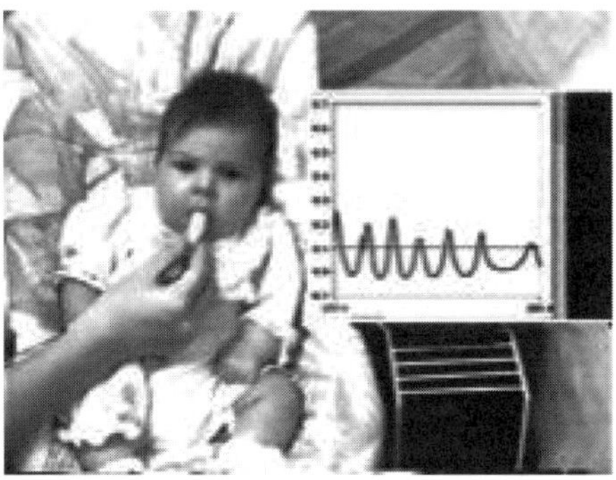

This picture shows a baby with a pacifier, but he is very unhappy. Look at his facial expressions, his eyebrows. There is no sound. Then, suddenly, the child hears sounds. The child freezes; there is much more concentration.

We found that the sucking response was markedly different in the analog condition compared with the nonanalog condition. In the analog condition, babies were more subdued in their oral response. They tried to explore their abilities to produce sounds that reflected their own proprioception.

One thing that is quite remarkable is that when the babies started to hear the sounds they produced, they tended to explore the threshold. Kids do not simply reproduce something that makes an effect — they discriminate the kind of effect that more or less reflects what type of action they are producing on the object.

What is interesting is that when we tried to reproduce or replicate this result with newborn 24-hour-old babies, who are very good at sucking and whose auditory system is fully developed immediately after birth, we could not find a differentiation between the analog and nonanalog conditions that exists in 2-month-old babies. This therefore represents a crucial development in terms of understanding their own bodies and activities in the world.

Another study shows that very young babies have a sense of their own bodies situated in the world. In this experiment, babies were presented with objects that were placed either just in reach or slightly out of reach by a few centimeters. The infants would then decide whether or not to reach for the object, depending on the perceived reach ability of the object to the body. We did a series of experiments that showed that by 4 months, babies have a very strong sense of what they can and cannot do by judging the world. They are situated with respect to objects in the world.

First Psychological Revolution: 2 Months

By 2 months, something very important happens: the emergence of smiling, the emergence of reciprocity. I would like to propose that this is the actual "psychological birth" of the baby.

This is not the biological birth, because we know that the first 6 weeks of a healthy infant's life are basically in perfect continuity with the behavior that the baby manifests during the last semester of pregnancy. Prechtl[e] demonstrated that there is a remarkable behavioral continuity between prenatal and early postnatal development.

[e] Prechtl HFR (ed.), *Continuity of Neural Functions from Prenatal to Postnatal Life*, Cambridge University Press, Cambridge, UK, 1991.

At 2 months, a profound change occurs: the emergence of smiling in a face-to-face interaction, as emphasized by Schaal, Ruff, and others. It is also, from the adult perspective, the time when parents and caretakers start to recognize a "person" in their baby, and not simply a "thing" that cries for their care. If you look at diaries of mothers and fathers, you will systematically find explanations or remarks with great emphasis on the discovery of the first socially elicited smile, not the reflex or automatic smile of the newborn. This is seen as an exchange in reciprocity, and is a major developmental step.

• • •

Snyder: That's interesting, but what about a study of breast-feeding and eye contact, the concentration of the baby on the mother's eye? This occurs even earlier than the emergence of smiling.

Rochat: Well, but there is no coconstruction of affects in breast-feeding. Actually, something very important does happen around the same age in the way children look at faces. By 6 weeks, when babies start to smile, there is a transition in behavioral organization at very basic levels such as visual attention. Children's visual attention is very different prior to and after 6 weeks.

You have to observe, to answer your question, how babies at 1 month versus 2 months attend to faces by looking at their eye movements, and seeing where in the face they look at. At 1 month, as you know, they focus on the high contrast outline and outside features of the face. At 2 months, they become much more interested in the internal features of the face.

• • •

Prior to 6 weeks, you can clearly see that babies' eyes are wondering in the periphery, not paying attention and not having much eye-to-eye contact, which is so crucial for reciprocity.

Reciprocity is expressed in early imitation. There are now a dozen experiments on babies showing that by 6 weeks, they tend to reproduce the facial gestures expressed by adults.

Meltzoff, a pioneer of this new wave of early imitation studies, published one study that I find particularly meaningful. He tested 6-week-old babies when facing an adult pulling his or her tongue out (i.e. tongue protrusion). The direction of the tongue protrusion varied: it was either at the midline or to the side (left or right) of the experimenter. What Meltzoff showed in his very minute and delicate analysis was that by 6 weeks, infants tended to reproduce not only the tongue protrusion as a whole, but also the tongue protrusion and the detail of its orientation. This indicates that children do not simply release a mechanism of reproduction, but demonstrate some rudimentary effort to match and explore the model and their own bodies.

• • •

Snyder: And that's at 6 weeks?

Rochat: Yes. Meltzoff did not try this experiment on newborns, but he was sure that it would work as well. However, I am sure that it will not work for newborns because, in my opinion, children only start to open up during the "2-month revolution."

• • •

What is interesting in terms of learning is reciprocity. You smile, I smile, and so there is this matching of emotional expression. We conducted an experiment in which babies were either facing themselves in front of a large TV screen or facing an adult experimenter who mimicked their movements. The experimenter did exactly what the baby did; for instance, if the baby clapped hands, the experimenter immediately clapped hands as well. We found that at 4 months of age, infants reacted very differently to their own images compared with the image of an imitating adult. They smiled more at adults, and vocalized much more at themselves. So, basically, they can make a discrimination between themselves and someone else.

• • •

Mogi: Did you measure the mean temporal delay in the others' condition?

Rochat: Yes, we used a time machine that allowed us the online delay that is used in broadcasting. Imagine a commentator who spills his coffee during a football game and cries, "Shit!" You have to cut that part away, so you need to buy seconds of delay. We did some kind of psychophysics of when the children started to disconnect with the visual figure. Typically, 3 seconds is the window around which they start to notice the delay. But, it's remarkable how tolerant to the delay babies can be.

Mogi: So, you mean a strong response to the self-image stops at 3 seconds?

Rochat: Yes, 3 seconds. But, this is a complicated issue because whether there is a delay or not is absolutely contingent. Every time I do something, the visual feedback does something. So, if there is a time delay, there is no coincidence over time, but there is always a coincidence over what is happening. It's complicated. To illustrate this point, we conducted an experiment whereby we borrowed a paradigm from Trevarthen, who did a study with Elizabeth Murry some time ago.

We observed a mother and her baby in separate rooms, and both of them communicated with each other live via a closed-circuit video system. In another condition, the baby was presented with a replay of the mother in a previous session. So, the mother looked happy, but was not attuned or sensitive to what the baby was doing. There was no feedback.

What Trevarthen and his collaborators found — and we somehow replicated — was that babies between 2 and 4 months started to manifest differential responses to their mothers that were either live or in playback. They showed different attitudes.

• • •

We did another experiment: the experimenters played a routine peek-a-boo game with babies. They were trained to do a certain number of gestures that fit the peek-a-boo game, which is a universal teaser to babies. Babies love it: "look-look-look, peek-a-boo, yeeees." It has this universal envelope from the crescendo of tension to the release of tension. But, in one condition, the experimenters heard instructions in the earphone to "scramble" the peek-a-boo game. They were told to deliberately disorganize the peek-a-boo game.

We tested 2-month-old, 4-month-old, and 6-month-old babies. At 2 months, babies responded equally to the scrambled and the organized peek-a-boo game. By 4 months, they responded significantly less to the scrambled game compared with the organized peek-a-boo game. This means that by 4 months, infants expect a certain form in the exchange or reciprocation of a dialog.

Second Psychological Revolution: 9 Months

Let me now come back to the central point of this talk. By 9 to 18 months, there is an emergence of coawareness. To illustrate this point, we borrowed Meltzoff's paradigm and expanded it a little for our experiment (Meltzoff is a very creative experimentalist).

Babies were faced with two experimenters. Both of them were of the same gender (female), looked the same, wore the same blue apron, and had the same swept-back hairstyle. Both of them had the same toy, which in this case was a little cube or a little ball. The baby sat on the mother's lap; and one of the experimenters was trained to imitate exactly what the baby was doing with the cube, while the other one did something that was not equivalent to what the baby was doing. So, again, here is an analog/nonanalog situation. We looked at what the baby did.

We found that by 14 months, babies paid much more attention to the person who was imitating them. They even tried to trick the person who was imitating them by engaging in an action, and then stopping suddenly to throw the imitator off balance. At 14 months, infants start to recognize themselves in others — others are becoming "social mirrors." The others are used to consolidate the first implicit, and eventually explicit, identity of the self.

Let me get back to my main point: self-consciousness is a bottleneck to creativity. We did a famous Piaget experiment: putting an (attractive) object on a blanket, and seeing when babies come up with the creative idea of pulling the blanket to get this object. Piaget claimed that this kind of pulling action emerges in 9-month-old babies. We tested the same task on babies aged 14 and 18 months old.

We found that by 9 months, 90% of babies pulled the blanket to get the object; however, by 18 months, the percentage was close to zero. Why didn't they do it? Because they knew that their mother was sitting near them. So, they saw the object and then went "ah, ah, ah, ah."

Of course, they could pull the blanket to get the object. They are perfectly capable of doing that. But, that is meaningless. What is important is not the physical task of getting this object; rather, the problem to be solved becomes social. The goal is collaboration and affiliation, not grabbing the object. Essentially, people, not the objects, become the focus of attention. This happens at around 14 months or later.

Conclusion

Since we started out by talking about the islands in the Pacific, let me just drop a few images here. We did all kinds of tests on babies in Samoa and Fiji. We looked at the mother in interaction, and tried to see if there were any intrinsic differences between what we observed in Atlanta (USA) and in the Pacific, where the culture is radically different. In the Pacific, there is no privacy, 40% of the kids are adopted, and kids are surrounded by very different sets of values.

What we observed was that at 14 months, babies requested for help when trying to open the box. Therefore, this behavior emerges at about the same age and is universal. I do not think that it is innate, but I think there are some basic competencies that come from social transactions.

So, this is the conclusion: self-awareness — the bottleneck to creativity — is not the awareness of the body, the ecological self, or what is situated in the world. It is essentially the "coawareness" or the awareness of the self as seen through the eyes of others.

I would like to finish with a piece of writing by one of my students. It is very validating for me as a teacher, but it is also very tragic. It is a tragic testimony of higher education in one of the most powerful (and currently damaged) countries, the USA. The student is very smart, a high-honors student.

> I have not had one serious discussion about creativity in a single class in my 3½ years in college. Not one. This class is my first exploration of creativity. And the insight I have gained is invaluable. I love the idea of the art of the drift [which is

something I developed]. People in general write or talk about the clear agenda, the goal. But, it is the very act of writing or talking, the process, that is critically important. This is a principle of creativity. The greatest discoveries were made by accident, not by intention. I, for one, am a believer in the process.

So, I find it quite extraordinary that higher education is devoted so much to the accumulation of knowledge; however, there is no reflection on what it takes to become creative. This workshop is really a great opportunity to reflect on creativity. Thank you very much.

Further Reading

Rochat P, *The Infant's World,* Harvard University Press, Cambridge, MA, 2001.

Baby Talk

Do you remember when your child, as a baby, began talking? Did he or she suddenly start to utter proper words? Probably not. According to Professor Nobuo Masataka, babies first start laughing. Newborn babies, however, cannot laugh well; they must practice to laugh longer.

Do you notice that babies always try to kick in the air while laughing? Professor Masataka says that babies kick because kicking is necessary to laugh rhythmically. This kind of kicking occurs at around 3 to 5 months of age.

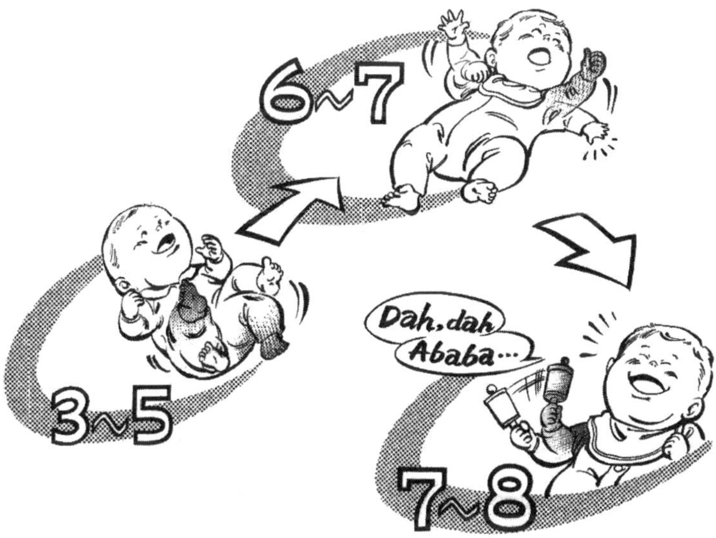

What happens next? If you observe babies carefully, from the age of 6 to 7 months, they try to "get the rhythm" by banging their hands. Babies bang hands not because they are angry; rather, they bang to laugh.

Of course, we adults do not know if babies are really happy when they kick and bang. But, their laughter gets better and better until, one day, the laughter changes into "ba, ba, ba . . . da, dah, dah" — (canonical) babbling. This babbling is, in a sense, the onset of language.

By the way, do you like Mozart? Some people say that Mozart is good for unborn babies. The following are sentences picked up, at random, from advertisements:

> "From the third trimester of pregnancy and beyond make music a part of your baby's life every day."
> "Listen, talk, and play music to your unborn baby!"
> "Mozart for mothers-to-be."
> "Gentle love songs for tender ears."

Apart from the scientific validity of prenatal music, newborn babies seem to love Mozart. Is this because of, for example, prenatal talks by mothers? In other words, are babies fond of Mozart because their mothers' talks remind them of Mozart? Or, is it something inherent in all of us? Are babies "brain-wired" to be fond of Mozart? Professor Masataka seems to have an answer to this question.

Babies born from hearing-impaired mothers have no chance of "listening" to prenatal talks by their mothers. So, if these babies show a tendency to love Mozart, we seem to be able to conclude that the Mozart-loving tendency comes from the environment. Mothers are responsible for this trait. On the other hand, if these babies prefer Mozart despite having no chance of listening to either Mozart or their mothers' prenatal talks, then we seem to be able to conclude that the Mozart-loving tendency is inherent in all of us (i.e. genetic).

Interestingly, experiments performed by Professor Masataka's team clearly show that the answer is the latter. In his experiment, babies showed more interest in the original Mozart music, and showed little interest in the tone-deaf or inharmonic version of the same piece.

We all love Mozart. This seems hard-wired. To be precise, Professor Masataka does not claim that Mozart's music is inherent. He thinks that human beings are hard-wired to love harmonic sounds and abhor inharmonic music.

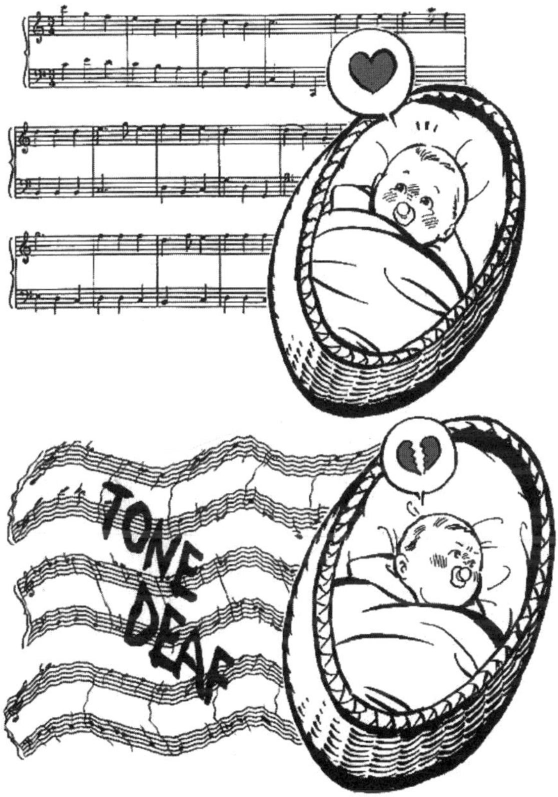

So much for the introduction. Please enjoy Professor Masataka's illuminating lecture.

The Onset of Language and Creativity

Nobuo Masataka is a professor at the Primate Research Institute, Kyoto University.

I am still wondering what I am going to talk about because my work has little to do with creativity. However, the time has come, so I have decided to present several findings from my previous work in language development with preverbal infants. These findings are independent of one another. I apologize in advance if these findings have little to do with creativity.

I am going to talk about several developments. First, I will discuss the motor learning aspect of early language acquisition, and then I will talk about the cognitive learning aspect of language development in preverbal infants. Lastly, I will show you the latest results of my noninvasive brain imaging studies, in which functional magnetic resonance imagine (fMRI) was performed on Japanese adults while undertaking a learning exercise.

Kicking and Babbling

Let me start by explaining motor learning in preverbal infants: how they acquire the ability to produce syllabic linguistic utterances, the so-called "babbling."

Previously, people believed that the onset of babbling was a consequence of the maturational process. However, this is totally misleading. Infants, including nonhearing-impaired infants, have tremendous difficulty in producing babbling. We cannot understand their difficulty, simply because they cannot speak or express it. We too had difficulty mastering babbling when we were babies; fortunately or unfortunately, we do not remember our infancy. Preverbal infants face the most difficulty in producing sounds with mouthy syllabic characteristics and consonant elements.

Usually, the onset of babbling comes between the age of 8 to 10 months. Before the onset of babbling, infants seldom produce mouthy syllabic utterances as linguistic production. However, they frequently produce another type of mouthy syllabic utterance as an emotional expression — laughter. Babies start laughing after the age of 3 to 8 months; and this ability is then expanded until the mouthy, syllabic, linguistic utterance of babbling appears. There is a close relationship between laughing and babbling.

Laughing is the species-specific pattern of emotional expression that we inherited from our ancestors — primates — and I happen to work for the Primate Research Institute. So, let me demonstrate a laugh-like vocalization of nonhuman primates. Even chimpanzees produce mouthy syllabic utterances, which we call the "pant-hoot." In fact, Charles Darwin noted that the pant-hoot is laugh-like, and argued that this is the prototype of human laughter. Have you heard the pant-hoot before? Probably not. Not many people are familiar with the pant-hoot. So, let me reproduce it by myself. Here we go: "Hoot, hoot ... hoh, hoh! Hoh, hoh, aghaaaa!"

There are generally three sections in the pant-hoot vocalization: "hohohoho," "howhowhowhow," and "augh augh augh."

Interestingly, we have evidence of the presence of dialects among chimpanzees. This is very interesting because there is very little evidence of the presence of dialects among nonhuman primates, although there are a number of studies on dialects among bids. However, the chimpanzees' pant-hoot shows geographical variation, so it is probably culturally inherited. For example, the figure below shows two different pant-hoot patterns for two different populations, which are geographically separated by just 60 kilometers. The most prominent difference is in the build-up section: Mahale chimpanzees show a very rapid build-up, "ho, ho, ho, ho!"; whereas JMC chimpanzees show a very slow build-up, "hogh, hogh, hogh, hogh, hough, hough."

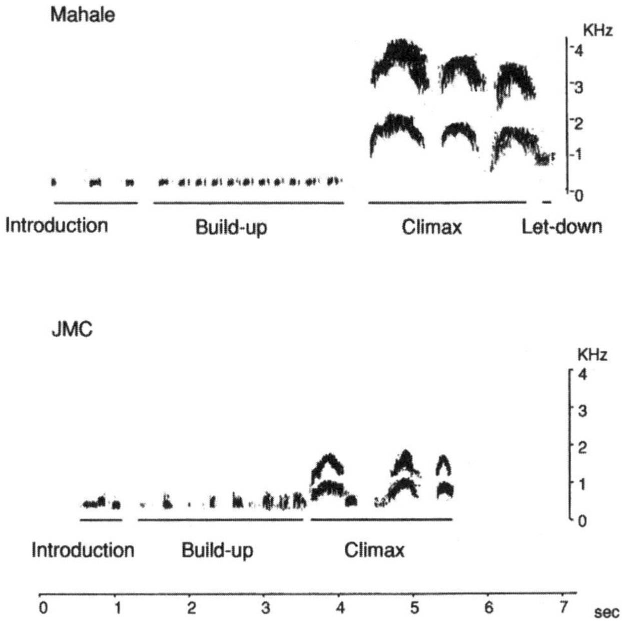

Representative sonograms of pant-hoot vocalizations of chimpanzees in two populations, Mahale and JMC.[a]

Chimpanzees need more than 8 to 10 years to be able to produce this type of vocalization completely. The difficulty lies in the build-up section: they have difficulty in repeating "hough, hough," i.e. in exhalation and inhalation. In contrast, when humans laugh, we only exhale; we do not produce "haugh!" while inhaling. This difference is due to the differences in vocal structure between humans and chimpanzees. Also, chimpanzees, unlike humans, do not produce "ha, ha, ha." When we produce this type of sound, we cannot speak. This is very important.

To be sure, newborn humans do show a chimpanzee-like pattern in the vocal apparatus. But, after birth, the vocal structure gradually changes. It takes 2 years for the change to be completed.

---

[a] Kajikawa S, Hasegawa T, "Acoustic variation of pant hoot calls by male chimpanzees: a playback experiment," *J Ethol* **18**: 133–139, 2000.

The first dramatic change is known to occur between the ages of 3 and 4 months. Babies begin to laugh, "ha, ha, ha, ha!" However, when babies express laughing of the emotional expression, at the age of 3 to 8 months, they do not only laugh. Laughing behavior is often accompanied by the rhythmic pattern of leg and hand movements: kicking and banging. This temporal co-occurrence is very important for the elaboration of mouthy, syllabic, vocal utterances.

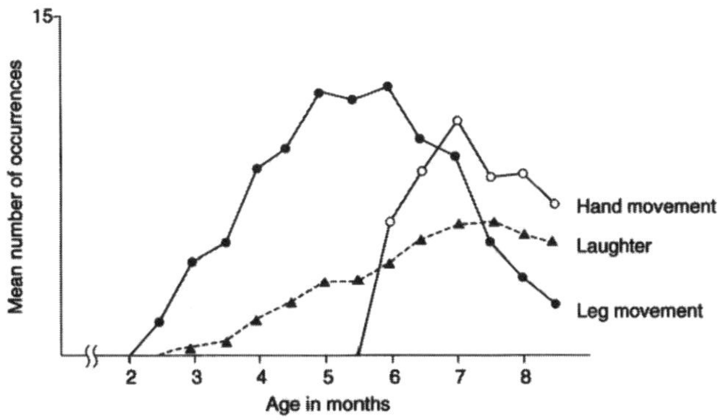

Overall developmental patterns in the occurrence of rhythmic leg and hand movements and laughter.

This figure shows the developmental profile of rhythmic motor movement and laughter, which first appears between the ages of 3 and 4 months. Rhythmic leg movement appears 1 month earlier, and is produced very frequently at the age of 6 months (the leg movement is known as the precursor behavior of bipedal walking: if you hold a 6-month-old infant upright, he or she will automatically produce a step-like walk at this age). Then, rhythmic hand movement appears; we call this "hand banging." Laughter shows a strong tendency of temporal co-occurrence, first with leg movement, then with hand movement.

What is characteristic of the first-phase laughter at the age of 3 to 4 months? Well, infants are not able to laugh for very long. They laugh only briefly: "Ha!" Laughing longer is very difficult for them because, to do so, they must keep subglottal air pressure (or vocal fold tension) constant while repetitively opening and closing the glottis. (This is the same difficulty that infants face when they try to produce babbling. They are unable to babble until the age of 7 or 8 months, simply because they are unable to perform this intricate movement.)

However, they "exercise" this movement through the temporal co-occurrence of laughter with stereotypical kicking, because to keep kicking longer is much easier than to keep laughing. So, generally, with temporal co-occurrence, they tend to laugh longer over time. For example, babies initially laugh twice while kicking five times. As infants grow older, laughing and kicking somehow become phase-locked; they are coincident with one another. Babies laugh five times and kick five times.

Then, the temporal co-occurrence of laughter with hand banging appears. Soon after this temporal co-occurrence, the behavioral pattern of laughter shows another (qualitative) change. Initially, infants are unable to laugh rapidly because it is very difficult to repeat the opening and closing of the glottis. However, with the temporal occurrence of hand banging, infants begin to laugh more and more rapidly. Then, at the age of 6 or 7 months, speech-like mouthy syllabic utterance — "marginal" babbling — appears. Marginal babbling is distinguished from canonical babbling.

• • •

Words: marginal vs. canonical babbling

> ... a pronunciation transcribable as "ta" would be deemed canonical if articulated with a rapid transition duration, but would be deemed noncanonical if articulated with a slow transition duration.[b]

• • •

Canonical babbling is characterized by the presence of distinctive consonant elements and mouthy syllables. Marginal babbling is ambiguous with regard to the presence of consonant elements. So, marginal babbling sounds like "a, a, a, a, a"; whereas canonical babbling sounds like "ba, ba, ba, ba, ba, ba."

---

[b] Masataka N, *The Onset of Language*, Cambridge University Press, Cambridge, UK, 2003, p. 169.

At the onset of marginal babbling, babies show a tendency for temporal co-occurrence with hand banging movements. Then, canonical babbling appears. Once canonical babbling appears, the production no longer co-occurs with rhythmic hand movements. Therefore, with such movements, infants probably come to acquire the ability to produce mouthy syllabic sounds with consonant elements.

To produce sounds with consonant elements, we must perform the movement of articulation very rapidly. For instance, in order for a sound to be perceived as having consonant elements (e.g. "da" or "ba"), we must perform the movement of articulation within 100 milliseconds. It should be very rapid.

Infants initially try to imitate the sound, but they cannot do it as rapidly as adults do. So, the sound cannot be perceived as having consonant elements by adults. Soon, however, with the temporal co-occurrence of hand banging, they acquire the ability to articulate rapidly.

So, here comes the first lesson drawn from motor learning. If a carekeeper is engaged with an infant in a very joyful or pleasant joint interaction, it induces the infant to laugh more frequently. Also, the infant must be facilitated to motor learning, as it is necessary for mouthy syllabic utterances.

• • •

Discussion

Audience: How do you know that the noise that infants make counts as laughter, as in adult humans?

Masataka: Oh! I conventionally call it "laughter." Whenever any rhythmic vocal production is recorded, we conventionally call it "laughter"; but we do not know whether the infant is really pleasant and joyful or not. Maybe I am preoccupied with this notion. However, in my opinion, people basically engage in joint production very playfully, and babies frequently laugh.

• • •

By 9 Months of Age, Babies Hear and Memorize Words

The age of 9 months is also very important for preverbal infants for cognitive development. Around this age, infants come to be able to memorize the words they hear. They can perform a long-term retention of heard auditory information. This was first experimentally demonstrated by Peter Jusczyk[c] of Johns Hopkins University. His article appeared recently, and we have confirmed his findings in various situations.

Let me describe our experiments. In the first experiment, we asked mothers of infants to read a picture book twice a day for a 2-week period. In the book, we prepared a target word as the keyword, which we ascertained was totally unfamiliar for the infant; for instance, *ringo*, which means "apple" in Japanese. So, infants were exposed to the word 28 times during the 2-week period.

During the following 2-week period, infants were not exposed to that particular sound. Then, we presented the target word and control words, and measured the duration of the infant's eyesight using the preferential looking procedure. Control words were prepared according to various conditions, such as phonologically similar words, (tango), words that were semantically close to the target word (*itigo* or "strawberry"), etc.

In any condition, infants showed a strong tendency to look at the target word (which was presented in the picture book) longer than the control words. So, around 8 or 9 months, they have a long-term retention of words that they hear if they are interested in the word.

At the age of 9 months, babies are able to not only memorize words, but also produce words by themselves. They are thus close to the emergence of postwords. In order for postwords to emerge, infants must assign the appropriate meaning to the words they have memorized. This is called the "mapping" or "categorization" program. How can babies categorize the appropriate meaning when they are exposed to a novel label to an external object? To investigate this, we did an experiment with 2-year-old children. Our conclusion is that to categorize the appropriate meaning for a memorized word, it is important for infants to observe the actions of the parents who teach the word.

---

[c] Jusczyk PW, Aslin RN, "Infants' detection of sound patterns of words in fluent speech," *Cognit Psychol* **29**: 1–23, 1995.

This can be demonstrated by a recent experiment. We showed infants a target object, which was an egg-like cube made of glass. We taught the children that this was "muta" ("muta" is a meaningless word). Then, when presenting the target, we performed two different actions. In one condition, we saw through the object to show the children that the object was transparent. In the other condition, we rolled the object while repeating, "This is muta, this is muta."

We then prepared a pair of test objects. One was egg-shaped and made of rubber; the other was pyramid-shaped and made of glass. The former object matched the target object in terms of shape, but not material. The latter matched the target in terms of material, but the shape was different. We showed these two pseudo-objects and asked the children, "Which is muta?"

The children were forced to choose one of them. When we emphasized the material aspect, children were more likely to choose the pyramid-shaped glass object. On the other hand, when we emphasized the shape of the target, children were more likely to choose the egg-shaped rubber object.

So, when children learn the meaning of a novel referent, they do not just hear and memorize the sound, but they also carefully watch the actions of the caretakers. This probably has something to do with the mirror system in the brain.

Brodmann Area 9

I will now talk about the result of a recent imaging study I conducted.

My English-speaking friends probably understand the meaning of roman numerals such as "V"; however, I have never seen a Japanese person who really understands the meaning of roman numerals. So, in our experiment, we asked Japanese subjects to figure out the meaning of roman numerals. Then, the first fMRI measurement was performed. After the measurement, we instructed subjects on how to read and comprehend roman numerals for 3 to 4 months, before performing successive fMRI measurements again.

Before the instruction, when subjects were totally ignorant of the meaning of roman numerals, the activation of vocalization was identified in the local areas of the brain. Once they were instructed on how to understand the numerals, the pattern of activation became more widespread, including a premotor area in the frontal cortex. For comparison, fMRI measurements were also performed during the intermediate and end stages of the learning.

Surprisingly, a similar pattern of brain activation was identified in the last three measurements. Furthermore, when we subtracted the activation of the intermediate stage from the initial stage, and the activation of the final stage from the intermediate stage, we found an increase of activation in a restricted single region: the left little dorsolateral prefrontal area, the Brodmann area 9. No decrease was found.

As far as we know, there are several studies that relate the learning changes of a practice to the changes in brain activation. However, all of these studies show a decrease of activation, probably because of the decreased dependency on the working memory system. In our case, we found an increase of activation! This is probably the first demonstration of an increase of brain activation with respect to practice-related changes.

Our results probably may not occur for language-learning practices, given that our experiment only involved learning to transcode roman numerals into numerical representation. However, for such transcoding, a specific region of the left prefrontal area of the brain may play a very important role.

In all, what I have drawn from both my own and others' conclusions is that various brain regions participate in the learning of languages. Of course, the local areas play a very important role. However, in addition to the local areas, other regions — including the premotor area, the parietal region (particularly the supramarginal gyrus), and certainly the prefrontal area — play a very important role as well.

If we take a look at each region independently, the functioning of each region can somehow be identified in nonhuman primates. For example, in monkey studies, we found the activation of each area with the use of similar paths. However, in humans, these areas relate to one another in a very coordinated pattern. This is probably very important for language learning.

Impairment and Genius

We also have evidence from sign language learning of the activation of the prefrontal area and the superior temporal gyrus in the mirror system. These areas work together to do language learning; and if there is an impairment in either brain region, infants have tremendous difficulty learning a language. The degree of language difficulty depends on where the impairment is.

What is interesting here is that the impairment is so restricted that we can still do pretty well; moreover, a very unusual or unique compensation appears. For instance, if we have an impairment in the parietal area, we have difficulty storing phonological information temporally, because the parietal region (submarginal gyrus) is very important for phonological grouping in the working memory system. So, we cannot store the information we hear temporally (apparently, Einstein had an impairment in this area). However, compensation occurs: people frequently become very visual. In some cases, they are able to store visual information temporally as we do for phonological information.

I claim that this is the basis of creativity. People do not see something as a whole; rather, they sometimes see each as a part, as Professor Snyder mentioned in his talk. In this way, something particular and very unique — a genius — may emerge. This is my conclusion. Thank you.

Further Reading

Masataka N, *The Onset of Language*, Cambridge University Press, Cambridge, UK, 2003.

Dr Jekyl, Mr Hyde, and Qualia

People tend to think that a brain functions like a computer. These two systems do seem to have many features in common. For example, they can process words, recognize visual patterns, produce sounds, do arithmetic, etc. There is, however, a crucial difference in terms of spontaneous activities.

Computers concentrate on doing the job. They work hard: they are workaholics. A spontaneous activity, which has little to do with the job, means "error," in which case the machine halts. Our brains, on the other hand, are full of spontaneous activities. So, when I watch Star Trek *on TV, my brain is not only performing that particular job (watching Star Trek), but is also doing all kinds of things in the brain's backyard. I do not know about these hidden activities going on inside my brain. After all, I did not order my brain to do these things. My brain seems to be spontaneously acting on its own, without my permission.*

Dr Mogi says that these spontaneous activities might be the source of our insight and creativity. If this is true, in order to build a "creative" or thinking computer, we must somehow implement these spontaneous activities into the computer architecture. At present, even a single spontaneous activity freezes the computer, so my favorite Mac is not really creative on its own.

Some of the concepts in contemporary brain science are hard to grasp, partly because the field of brain science is still in its infancy.

New discoveries keep flooding in. For example, Dr Mogi mentions the "preferred cortical state" in his lecture. This is an activity state of a pack of neurons where the activity is at its maximum. The brain seems to be spontaneously browsing through many states, probably getting ready for action. The true function(s) of this state, however, is still under debate.

Dr Mogi's main theme of interest is "qualia." What are qualia? Qualia are the hallmark of our subjective experiences, such as the "redness" of blood when you cut your finger, the "itchiness" when you scratch your skin, or anything else you personally experience.

In physics, the color "red" is explained as the state of an electromagnetic wave that has a wavelength of approximately 700 nanometers or less than one millionth of a meter. However, I personally do not see or feel that number. I only feel "redness." Nothing more, nothing less; just the subjective feeling of "redness."

So, physics does not explain our subjective feeling of "redness" or qualia. Can brain science (e.g. brain physiology) explain qualia? The answer is not yet. Qualia are there. We all know that. Qualia are our experiences. But, it remains one of the biggest mysteries of modern science.

Dr Mogi stresses the role of "emotion" in brain science. One example is "role playing." This is actually nothing special. We do it all the time. You think you are "you," when in fact there are many "you's" acting differently in society or in the human network.

For instance, I always eat lunch at a small restaurant near the corner of Peel Street. I always sit on the same seat. The same waiter comes and says, "Hi, a black coffee and today's lunch?" I say, "Yep, nice day, isn't it?" (If the weather is bad, I will say, "Yep, terribly cold, isn't it?" or "Yep, a soaking downpour, achoo!")

One day, I was walking down the street. A stranger passed by me and said, "Hello!" It took me several seconds before I recognized the man. He was the familiar waiter of the restaurant.

I could not recognize him because his "role" was different. Usually, he plays the role of a waiter. I encountered his new role as a pedestrian. He was dressed differently, and he even walked with a different rhythm.

In a sense, we all have a Dr Jekyl and a Mr Hyde — and many other personalities — living inside our brain. Of course, Dr Jekyl and Mr Hyde have totally different emotional behaviors.

Now, what do all these have to do with creativity? It's all yours to be discovered in the following lecture.

What Is the Difference Between Your Brain and a Computer?

Ken Mogi is a senior researcher at Sony Computer Science Laboratories, Inc., Japan. He is a visiting professor at the Tokyo Institute of Technology.

Many people compare computers with the brain. The brain is creative, whereas the computer is not creative. Why? The most obvious difference between these two systems is that in the brain, there is a high level of spontaneous activity of its units or neurons.

I once studied the neurons in the middle temporal area (MT), which is responsible for motion perception. I found out that only 30% of the neural activity can be explained in terms of stimulus property. The remaining 70% are the "noise" in the conventional sense. On the other hand, the computer shows little spontaneous activity. If there were even a small percentage of spontaneous activities, let's say 0.1%, then the computer would malfunction. It would simply stop functioning. So, this is the most obvious difference between these two systems.

Neuroscientists do not know the functional significance of these spontaneous activities. For example, they have recently discovered that neurons in the primary visual cortex have a "preferred cortical state." This means that, even in the spontaneous activity state, the neurons correspond with the functional state in the evoked condition, i.e. with the external stimulus input. So, even when the brain has no input, it is still active on its own. Neurons are similarly activated in both the functional state and the preferred cortical state. So, spontaneous activity seems to be a very fundamental aspect of the brain function.

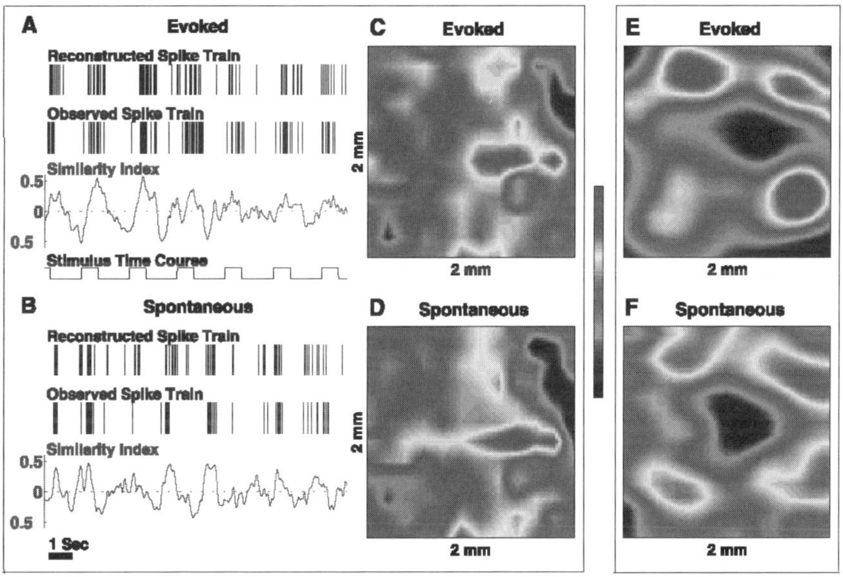

You are probably aware of the experiments by Valera *et al.* They found out that when you look at a black and white figure and suddenly realize there is a face in the figure, there is a very wide spread of synchronous activity among all the neurons in the brain. Many people suspect that, although there is no hard evidence, this kind of discovery can only be explained in terms of the spontaneous activity in the brain.

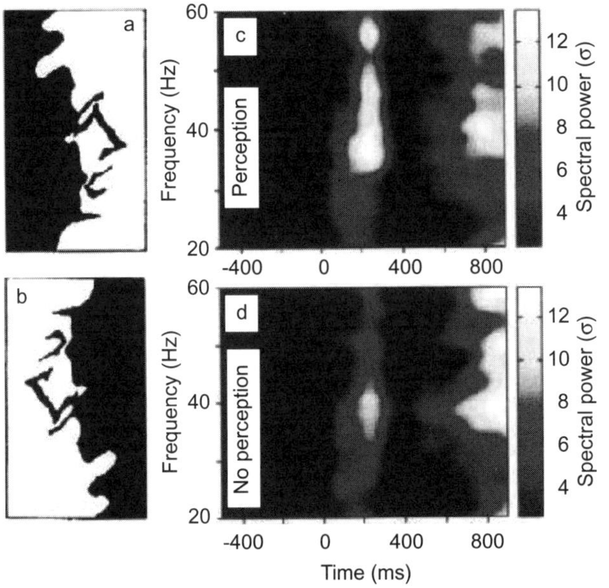

Valera *et al.*

We Need a 'Virtual' Homunculus

In line with Dr Kitano's presentation, I suspect that there is a genuine problem in striking the balance between evolvability and stability. You would like the brain system to be dynamically changing; at the same time, you want to have a stable perception system. So, if you see something red, the neural activities behind that perception might be different from time to time. Yet, you see "red" as identical to the same "red" that you saw in your first experience of the color.

I suspect that qualia have something to do with providing the perceptual stability, which is a necessary condition for a very dynamic and spontaneous brain system to be functional. (Maybe computers lack this aspect. It would explain why they cannot function when there is even a small level of spontaneous activity.)

On our first day in the neuroscience classroom, we learn that "there is no homunculus in the brain." In other words, there is no little man observing the neural activities in the front of the brain.

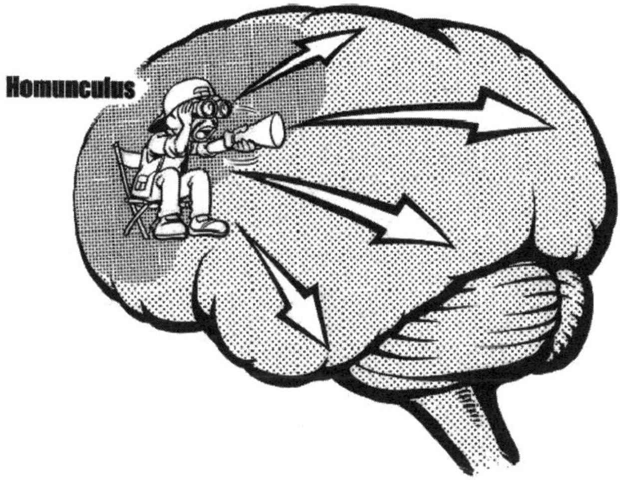

Of course, this is a very ridiculous idea because this kind of scheme immediately leads to an infinite regress.

However, in this presentation, I argue that we do need to have a "virtual homunculus." We do have a need for such a mechanism to integrate the parallel processes in the brain. We need the framework of the cell. This is a necessary condition for a dynamical system like the brain to be functional.

So, I would like to argue that qualia provide a necessary perceptual stability on top of this kind of integrative mechanism, beyond this kind of simplistic, classic, homunculus model. Then, in the latter half of my talk, I would like to discuss the role played by the emotional system: how these three elements — qualia, virtual homunculus, and emotion — combine together to help us to be creative.

Even no-nonsense, practical scientists like Crick and Koch talk about neural correlates. In their paper published in 2003, they argued that although the hypothesis of the homunculus is out of fashion, it is exactly how we all feel. It fits nicely with our subjective experiences. If something is an overwhelming illusion, it must in some way reflect the general organization of the brain.

> The hypothesis of the homunculus is very much out of fashion these days, but this is, after all, how everyone thinks of themselves. It would be surprising if this

overwhelming illusion did not reflect in some way the general organization of the brain.[a]

Now, for those of us who are interested in conscious experience, there is no denying that there is a huge difference between the states you are in when you are sleeping and the states you are in when you are awake. People who are interested in the question of consciousness are currently working on the "neural correlates" concept. There are billions of neurons activated in the brain, and there is the problem of qualia, so maybe the manifold of qualia is subjectivity. We are studying this correlation between neural correlates and conscious experience.

Neural correlates of conscious experience (qualia)

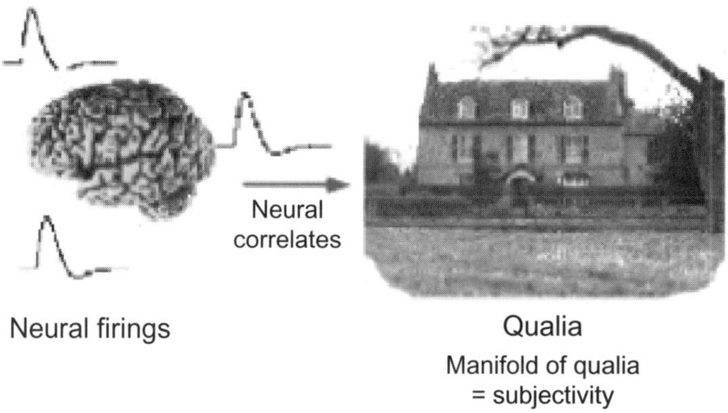

On the surface, the "neural correlates of conscious experience" concept seems to be free from the homunculus fallacy, because we are only talking about mapping the physical states and conscious phenomenal states. There is no need for any homunculus because we are talking only about mapping (this is what Crick and Koch say).

However, I have very strong reservations about whether the neural correlates concept really exorcises the homunculus. I do not think it does. I think that there is a genuine need to construct the nontrivial models of subjectivity.

Why do we have this illusion of the homunculus? Well, first of all, visual awareness is massively parallel, as I will now demonstrate.

[a] Crick F, Koch C, "A framework for consciousness," *Nat Neurosci* **6**: 119–126, 2003.

This is a very classic figure, so some of you might be familiar with it.

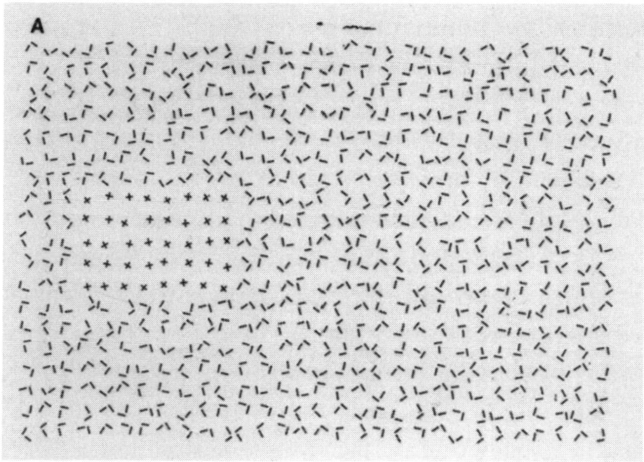

The region in the left of the figure immediately pops up. But, there is a second region which is distinct from the other regions in that there are "T's," as indicated in the rectangle below.

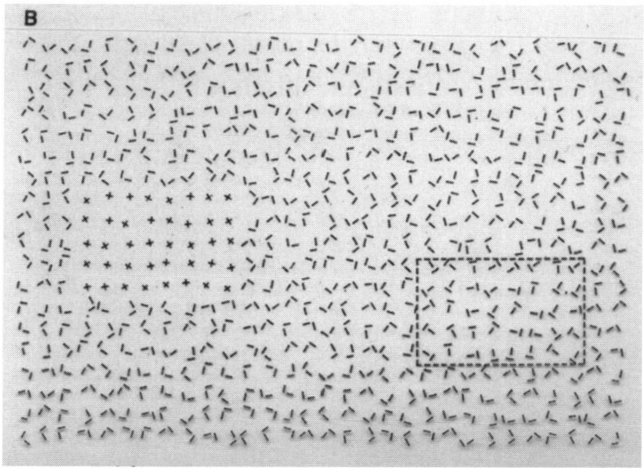

In order to make this second region distinct, you need to go through the symbols one by one. These are called "second-order differences," so the region does not immediately pop up. It so happens that this one-by-one practice is, so to speak, serial; but on the other hand, this is a visual perception. So, it is a massively parallel process. This massively parallel nature of visual awareness is something that needs to be explained when trying to pin down the brain function.

I will give you another example taken from a book on Goedel's theorem by Ernst Nagel and James Newman (1958).[b] Let's say there are five properties about the set K and L:

1. Any two members of K are contained in just one member of L.
2. No member of K is contained in more than two members of L.
3. The members of K are not all contained in a single member of L.
4. Any two members of L are contained in just one member of K.
5. No member of L is contained in more than two members of K.

The problem is the following: prove that K is contained in just three members.

Now, this might be a very interesting logical exercise. Professor Snyder might be able to solve it in 1 minute, but this kind of problem is usually rather difficult to solve quickly.

A computer using a formal system would solve this problem in a serial, step-by-step way. However, the moment we arrive at this interpretation of the postulates, we immediately know that the conjecture is true.

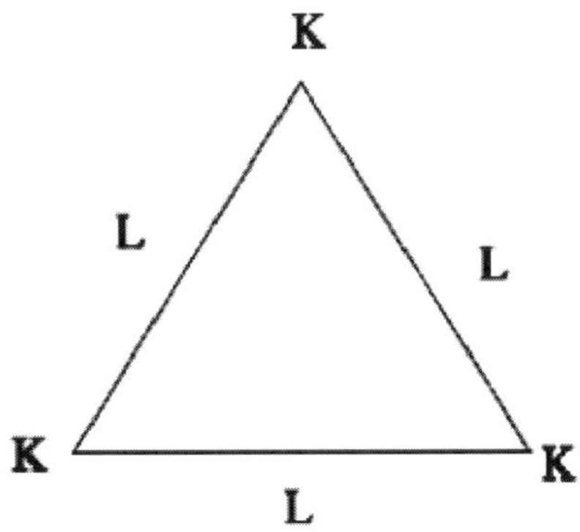

So, again, we face the situation where the massively parallel nature of visual perception is at work.

[b] Nagel E, Newman JR, *Gödel's Proof,* New York University Press, New York, 1958.

This is something that really needs to be done when you want to pin down what constitutes creativity. When you have an insight, when you are creative, you always have this kind of pattern recognition process. So, we really need to explain how the brain does it — this kind of massively parallel visual perception.

To summarize what I have been arguing, the "virtual homunculus," though we still do not know how it is constructed, perceives qualia — the perceptually stable element — in a massively parallel manner. I call this "integrated parallelism"; and this is at the center of the brain function, something that no computer can do at present.

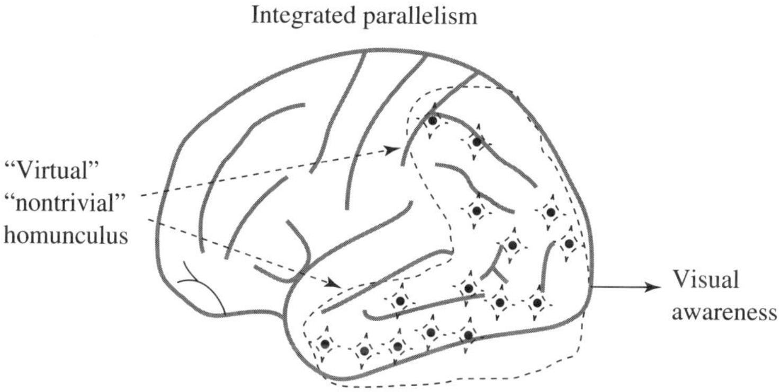

"I" see many things in the visual field in a massively parallel manner.

Now, many experiments have suggested that the center of the "self" is in the prefrontal region (e.g. area 46). So, we are finally talking about visual awareness. We are talking about a mechanism that makes these clusters of neurons accessible as a collection of qualia, from the vantage point of this region. This therefore means that there must be some interaction between the widely separated brain areas.

As an example, I have been doing an experiment on binocular rivalry. It seems that in order for us to see something, we do need to have a match between the bottom-up sensory process and the top-down intentional process.

So far, we do not know the nature of these matching processes, but I think the old ideas such as "neurons of area V4 representing color" no longer apply. This is because in order for us to see something as "red," we need not just new activities in area V4, but also a match between the two processes — hence, the very massively parallel nature of visual awareness. Without this aspect, we cannot really talk about information or representation in the brain.

One of the key elements here is time. It seems that there are at least two elements of psychological time in relation to the physical time in the brain. Maybe this sensory simultaneity corresponds with what Professor Poeppel referred to as the "smallest element of temporal moment," i.e. 30 to 40 milliseconds: the passage of 30 to 40 milliseconds is a kind of "freeze" in your time to a moment in psychological time.

It seems that there is a top-down arrangement of the different elements of perception. In general, there is a genuine problem: how to assign simultaneity to elements of perception that have different durations of processing time. For example, it is known that in order to process the form of objects, which needs to be processed up to the temporal cortex, it takes some hundred milliseconds. But, on the other hand, it takes just 50 milliseconds to process color. So, if this kind of particular physical stimulus comes up to area V4, the physical time required for the processing of color and form are different; yet, we perceive these as assigned to the same psychological time.

So, there should be an active, intentional process that assigns the perception with different physical processing times as simultaneous in psychological time.

When you tap something, you feel that the tapping sound occurs simultaneously with your action. This is because the brain actually perceives your action (which occurs prior to the sound perception) as somewhat later than the actual physical time. The tone — this "bang, bang, bang" sound — is perceived as somewhat later than the actual physical time so that, by this differential processing of the assignment of psychological time, you feel that the sound produced by your tapping is simultaneous with your action.

On the other hand, by applying transcranial magnetic stimulation (TMS), you can evoke involuntary stretching of your muscles. This is perceived as occurring earlier than the actual time. The tone is now perceived as occurring later than the actual time.

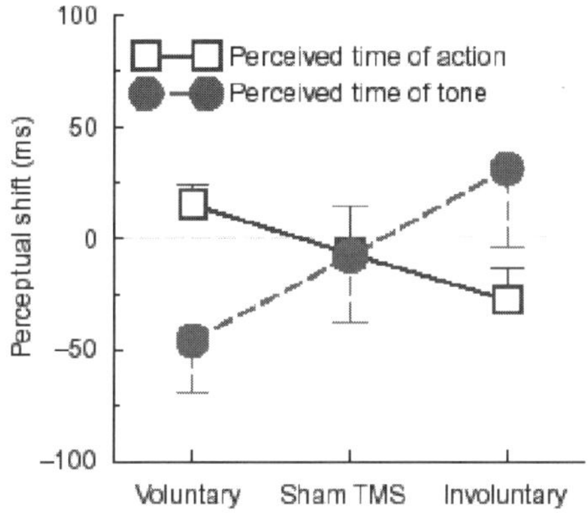

Haggard *et al.* (2002)[c]

So, the actual or physical time is totally different from your psychological time. Your brain does the fine-tuning. This kind of book-keeping — trying to make things consistent — seems to be behind our perception of sensory motor coordination and our perception of the world as a coherent structure.

So, as Professor Poeppel argued in his presentation, the construction of psychological time is no trivial matter. In my opinion, this is one of the key elements of what we call the "homunculus illusion."

Now, I'd like to argue that in order to explain how the brain achieves this fusion, we need to seriously think about the metacognition process. This is Professor Rochat's area of expertise.

It is more or less established that there is a "theory of mind": the perception of the interior mind and the perception of others' minds. This metacognition of others' minds and your mind occurs almost at the same time at around 4 years of age. So, metacognition (of my mind or of others' minds) seems to be the key element of social cognition.

Mirror neurons are another key element of this process. Recently, mirror neurons have been generalized to the consistency relationship between sensory and motor processing. For instance, if you make a sound by tearing a sheet of paper, the mirror neurons respond to the action of tearing the paper and also to the sound of the paper. So, I think the name "mirror neurons" might be a misnomer.

[c] Haggard P, Clark S, Kalogeras J, "Voluntary action and consious awareness," *Nat Neurosci* **5**: 382–385, 2002.

These neurons are responsible for making a consistent interpretation about the world, i.e. a contingency interpretation. So, when you perceive that others' actions are somehow ready for your own action, it just means that these two actions are contingent.

• • •

The origin of consciousness, I think, is the single most important difficult issue to be solved intellectually.

For me, all other problems are just details. If you take the real physicalism or functionalism, then you can of course explain the function of the brain, as Professor Poeppel nicely stated. There is no need for such terms as "emotion" or "perception" because it is just the physical property of the brain.

In contrast, the single most unexplainable fact is that we have conscious experience, which we cannot explain in terms of physicalism or functionalism. I am very much interested in explaining this. So far, nobody has had success.

Metacognitive Homunculus Model

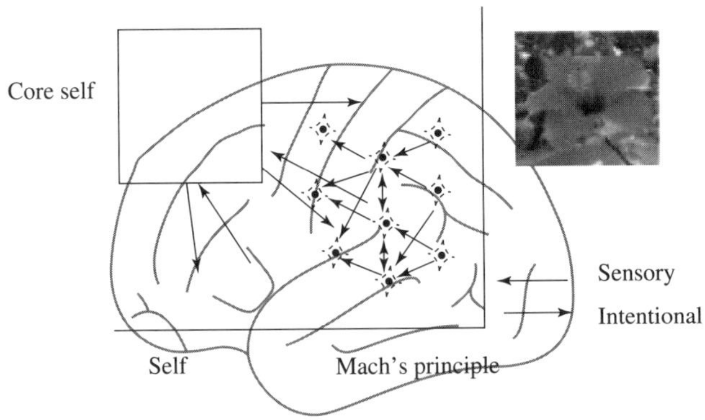

Metacognition of the self by the core self generates qualia.

• • •

Talking About Dopamine

The second half of my talk is about emotion. I have been enthusiastic about the subject of emotion for the past 6 months. I have realized that there is a completely new type of research area concerning emotion.

Emotion is ubiquitous in cognition. Why is that? It is because we are living in an environment that is uncertain, and yet we have to make decisions and take actions.

Daniel Kahneman studied the area of behavioral economics, which is the study of judgment and uncertainty. People have been talking about making a connection between this field and brain science.

Now, what is uncertainty? It is the reward-related area in the brain. I am talking about dopamine neurons here.

The old idea of reward goes like this: food is the reward; water is the reward; sex is the reward. So, biological systems favor reward and avoid punishment. This very simplistic idea, however, is no longer tenable because people (e.g. in the reinforcement learning paradigm) have been talking about "reward functions" or "reward probabilities" (in economics, they have been talking about "utility functions"). This means that if you take this action, you are given this reward with this probability, and so on. But in many cases, in real-life situations, this kind of reward function is not easily known; there is not even a probability distribution. So, the reward system must function in this kind of condition.

How do biological systems do it? It has been found by Schultz and his colleagues in Cambridge that an appropriately presented uncertainty can be a source of secondary reward. This is a very interesting discovery. I predict that this will be regarded as a mirror-neuron discovery in the field of emotion.

Why is this discovery so important? Now, Fiorillo *et al.* (2003)[d] studied the functions of dopamine neurons in the brains of rats and monkeys under three conditions:

Condition 1: Given the stimulus, they were given the reward with a probability of 1; so, they were always rewarded.
Condition 2: They were rewarded with a probability of 0.5, so they were rewarded once every two times.
Condition 3: They were not rewarded.

[d] Fiorillo CD, Tobler PN, Schultz W, "Discrete coding of reward probability and uncertainty by dopamine neurons," *Science* **299**: 1898–1902, 2003.

They found that there was a very prolonged dopamine neuron activity when they gave the monkey a reward with a probability of 0.5, i.e. when uncertainty was maximum.

Fiorillo *et al.* (2003)

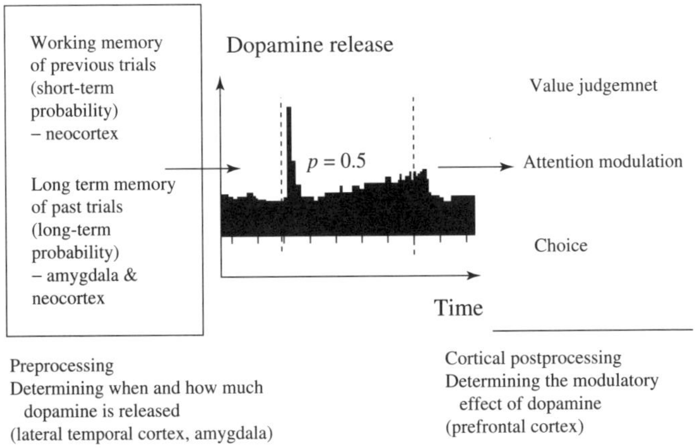

So, it seems that dopamine neurons code uncertainty as a kind of secondary source of reward.

When you think about it, when, how, and how much your brain releases dopamine are the most precious questions in life. We can give a talk about the new field of "dopamine management." For example, alcoholics release dopamine when they drink alcohol; they are addicted to it. You guys are addicted to research; you guys are addicted to finding something new. Your brain does it in a very uncertain condition. If you study very hard for your whole life, either you might not find anything new or you might win a Nobel Prize. We do something anyway, even if the reward is not known.

So, there is this problem of determining how much and when dopamine is released, as well as how the dopamine release affects your choices (perceptual judgment, attention modulation, and choice). There is this whole rich area of how to handle uncertainty in a robust way. This is the core of the emotional system. Therefore, emotion is not something primitive; rather, it is something that requires the highest intelligence.

You need a subtle balance between exploration and exploitation. A living system that just explores for an uncertain reward would hedge, because it is unlikely that it would get the reward. On the other hand, a living system that just exploits an already known reward would not be very expansive; it would not find a new rich ecological system. So, you really need to strike a balance between exploration and exploitation. This is what the dopamine system and other elements in the emotion system are doing.

In developmental psychology, John Bowlby (1951)[e] has talked about the necessity of the mother or carekeeper in general, providing a safe protection base for the infant to explore. For example, Professor Masataka talked about the preferential looking paradigm. According to this paradigm, it is known that infants look longer at a novel stimulus, but this only happens when the infant is safely protected by the mother's love. So, the perception of a psychological safe place or attachment to a carekeeper is a necessary condition for the infant to be explorative. This is well known in developmental psychology, of course. It also has a lot to do with the functional emotion system that I have been talking about.

In my view, another very important aspect is that when we talk about reward, we tend to be overly concerned with the "external" reward function. External reward means that you are given food, water, or an electric shock, for instance. However, it seems that there is this very rich ecological system within the brain: "internal" reward production and consumption, so to speak.

In Japan, there is a very strange group of people called *Otaku* ("nerds"). They are obsessed with *anime* ("animation") characters. They even say that *anime* female characters are better than real human females. Naturally, Japanese girls do not like these *Otaku* boys, so *Otaku* boys have not reproduced well. But, the fact that they keep existing tells you that the brain can make up its own reward system. In the case of *Otaku*, although they do not reproduce well, there should be a neutral property. No matter what kind of reward system you may have, it does not really matter for your survival.

I would like to mention the neutral theory of molecular evolution, which was conceived by Motoo Kimura (1968).[f] He established that a genetic sequence evolves faster when the sequence does not have any functional significance. This argument was counterintuitive at the time,

---

[e] Bowlby J, *Maternal Care and Mental Health*, World Health Organization, Geneva, 1951.

[f] Kimura M, "Evolutionary rate at the molecular level," *Nature* **217**: 624–626, 1968.

but it has since been confirmed. Kimura's theory is the foundation of the molecular core, by which you measure evolutionary distance.

In the case of the emotional system in the brain, we can think of a "neutral theory of emotional evolution" in the sense that we always experience emotional ups and downs, and that it evolves fastest when it has no survival significance. Maybe the *Otaku* culture is one demonstration of this (of course, this is pure speculation).

In my lab, we have been studying what we call "flat reward condition": no matter what we do, the reward is always the same. For example, when you buy a lottery ticket, no matter what number you choose, the reward expectation is the same; but, you are still very much concerned with the number you choose or where you buy the lottery. Human behavior is very much influenced by this kind of prejudice.

To illustrate this point, we conducted a gambling experiment. Based on the throw of a dice, you could either bet and win or bet and lose; if you did not feel like betting, you could pass. We found out that, based on the previous state (win, lose, or pass), the probability of betting changed without subjective noise. This is very strange because, mathematically speaking, no matter what the previous outcome is, it does not really affect your present probability of winning; and yet human behavior is affected by the previous results. There is some law here. This kind of behavior can only be explained by the function of the internal reward system.

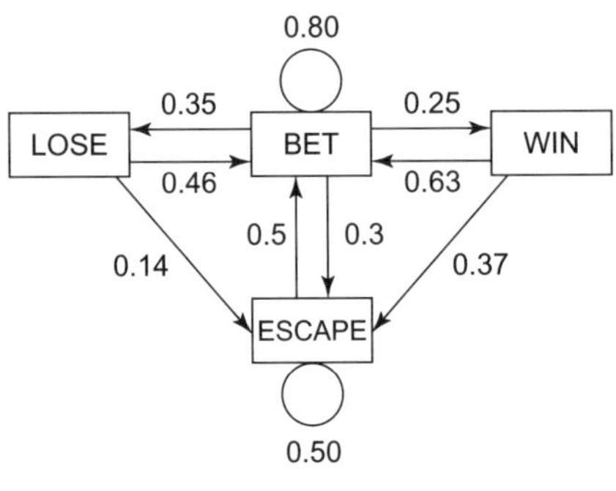

`Onzo and Mogi (2005)[g]

[g] Onzo A, Mogi K, "Dynamics of betting behavior under flat reward condition," *Int J Neural Syst* **15**: 93–99, 2005.

We have been trying to find out the temporal factor involved in this kind of effect, and we found that this effect lasts very long. If you compare this behavioral data with the neurochemical temporal factor in the emotional system in the brain, then we are on the right track in the field of neuroeconomics.

By the way, "neuroeconomics" is probably a misnomer because it sounds like people have a chance to make money out of brain science. Instead, what we are concerned with here is how the brain tries to handle uncertainty in a robust way. This is a very fundamental issue.

I think, in general, emotion has very much to do with creativity. Actually, when Affenhee, a Russian, came to Sony Computer Science Lab 2 years ago, his talk really inspired us. After the talk, I drew up this diagram:

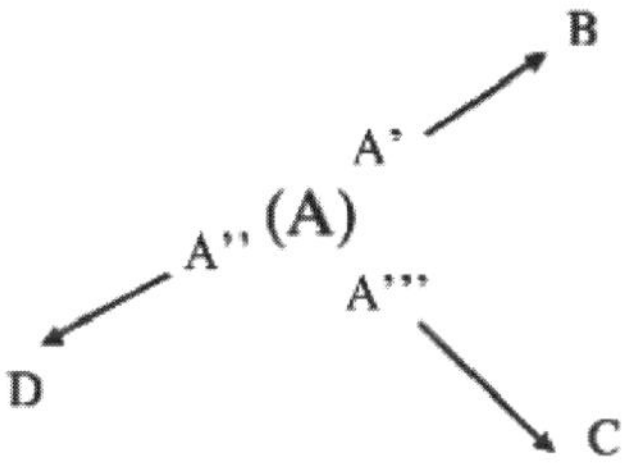

In Japan, for example, when you ride a taxi, you tend to engage in a certain conversation with the taxi driver. You might talk about whether the Tokyo Giants (a baseball team) won last night, or you might talk about the economy. But, you never talk about these things in daily life. When you are in a taxi cab with the driver, you are in a very strange mode of self. I do not know if this is a ubiquitous, universal phenomenon (e.g. in Germany or in Australia), but it is part of the Japanese psyche to carry up a certain typical conversation in the taxi.

So, depending on whom you interact with, you generate to define the self. When you talk with Mr C, you behave as Mr A'''. A taxi driver Mr B changes you into Mr A', and so on. Your self — Mr A — changes.

Recently, there has been a very dramatic demonstration of this kind of brain mechanism in the case of multiple personality disorder: dissociated identity disorder. No matter how many personalities you have (like Billy Milligan, you might have 24 characters or personalities), these personalities are divided into two categories: one has access to traumatic childhood memories of youth, and the other does not have access to traumatic personality childhood memories.

This experiment was conducted in Sweden, using people who could voluntarily switch between the two different categories of personalities. The researchers found that there is a particular area in the right prefrontal lobe that is responsible for suppressing traumatic childhood memories.

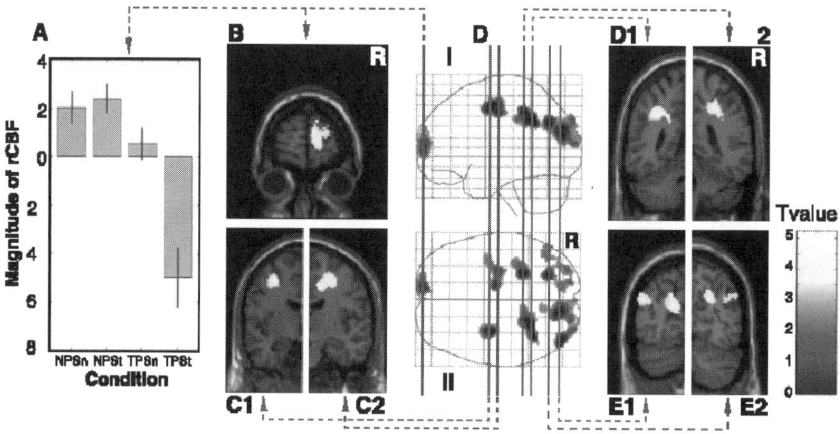

So, we can never talk about character memory or personality memory. When you go back to your hometown, you enter a particular personality mode. When you are at home with your wife and kids, you are in a particular personality mode. It is very likely that the brain (especially the prefrontal region), by controlling what kind of memory you access at that time, sets up the optimum personality trait, so to speak.

I believe this kind of mechanism is very fundamental when we think about creativity. Even scientists, when they try to think about new things, might unconsciously be envisioning themselves as communicating that discovery later to their colleagues or companions. If there were no scientific colleagues, would you really work very hard to come up with a new scientific discovery? I very much doubt it. So, this kind of mechanism (although it is a very extreme case) might be fundamental in selling out different selves.

Another example of the role of emotion in creativity is the Capgras delusion. The Capgras delusion has a lot to do with emotion because, say when you recognize somebody, your usual emotional quotation of familiarity is lost. You make up all these fantasies (e.g. this guy is not really Kaoru Takeuchi, but is actually the latest humanoid from Sony or a magician pretending to be Kaoru Takeuchi). All these are strange stories. This kind of phenomenon is documented in the book *Phantoms in the Brain* by Ramachandran and Blakeslee (1998).[h]

Of course, this is very strange, but you could also say that people are being creative in these situations. This is because in the normal processing of their daily lives, they would never dream of him as the latest humanoid or whatever. It is the disruption of the emotional system that creates these illusions.

I think this is a very clear testament of the fact that the emotional system is behind our creation of many things. You could even say that creativity is a reaction to the uncertain world by the emotion system.

Have you heard about the "leaping" penguins metaphor? The first penguin to leap into the water is very courageous because it does not know if there is a predator in the water. So, when you go to the zoo, you may see penguins hesitating to enter the water, with each penguin wanting others to jump into the water first to make sure that it is safe. But, if nobody jumps into the water, then you cannot get proof. Therefore, in spite of uncertainty, we sometimes endanger ourselves and jump into the water.

[h] Ramachandran VS, Blakeslee S, *Phantoms in the Brain*, William Morrow, New York, 1998.

To sum up, my presentation is based on two theses. One is that the main aspect of perception is how the brain provides perceptual stability in terms of qualia on top of spontaneous neural activity in the brain, and how the virtual homunculus processes the world in a massively parallel way. On the other hand, you have this emotional system trying to deal with uncertainty in the environment in a robust way.

• • •

Discussion

Tokoro: I have difficulty understanding what you've said, especially the relationship between uncertainty and creativity, and uncertainty according to the example that you have shown us.

Mogi: Internal spontaneous neural activity can also be social internal uncertainty. In order for the computer to come up with a new conception or representation, it has to have a spontaneous activity of its unit in a nontrivial way (not as a noise) that you know of as something that can be incorporated in the semantic system. I might conjecture that internal spontaneous activity is the essential element of the brain mechanism for handling external uncertainty. So, internal uncertainty and external uncertainty have coevolved to generate new ideas or concepts. This is lacking from today's computing machine.

Tokoro: And internal uncertainty comes from noise?

Mogi: Well, when you say noise, you immediately think of white noise, a random noise. But, as Professor Poeppel nicely put it, you can forward synaptic connection you get from any unit to any unit. So, given this physical reality in the brain (each unit needs to carry multiple contents), the brain neurons in the MT area not only process emotional information, but are also influenced by other emotions, value judgments, attention, and so on. So, the concept of noise has been analyzed with just one context in mind. A neural activity is not simply a noise; it's something that reflects and carries out multiple contexts.

Rochat: In the context of uncertainty, do you think superstition is a part of creativity?

Mogi: I think so, yet . . . take the film *A Beautiful Mind*, for instance. In the film, based on a true story, Nash became schizophrenic and had an illusion of a girl. This girl was a prototype of his creativity, yet Nash realized by logical reasoning that this girl was not real. The reasoning was that he had been seeing this girl for 10 years, but the girl stayed the same age for the whole 10 years. So, Nash realized that this girl must be an illusion.

Computer and Creativity

Geniuses and People Around Them: The 'Dark Side' of Science?

Luc Steels is a professor of computer science at the Free University of Brussels (VUB); founder and director of the VUB Artificial Intelligence Laboratory; and cofounder and former chairman of the Computer Science Department, Faculty of Sciences, VUB. He is currently the director of the Sony Computer Science Laboratory in Paris, where he leads a team of prominent researchers in fields as diverse as computational linguistics, music perception and processing, artificial intelligence, robotics, brain science, and interactive technologies.

Steels' scientific research interests cover the whole field of artificial intelligence, including natural language, vision, robot behavior, learning, cognitive architecture, and knowledge representation. Recently, his focus has been on dialogs for humanoid robots and on fundamental research into the origins of language and meaning. In the well-known "talking heads" experiment, Steels investigated how language emerges through the interaction between agents. He has written many scientific papers and books, including one coauthored with Rodney Brooks. Steels is known among his colleagues for his soft-spokenness, good sense of humor, and artistic sense in presenting his works both for the experts and the general public.

I would like to follow up on Dr Mogi's challenge to think about genius and creativity, and to consider the aspects that we recognize in highly successful, creative people. Some of them have been discussed in this workshop.

Now, I am going to mention some famous people that everybody would agree are geniuses. The first thing I would like to draw your attention to is the cognitive and personality factor, in particular the social aspects.

If you talk about Marie Curie, you also talk about Pierre. They were together, and he was killed in a car accident. His daughter was also killed in a car accident. So, in this context, if you were to take the same person and move him or her to another context, nothing would happen. So, the context — the interaction with the others — is important.

Virginia Woolf was an author. People usually think that authors are very lonely people. However, Woolf was actually very much part of the Bloomsbury Group, and she published her books together with her husband Leonard. She was also very critical of the fact that there were so few women who had won the Nobel Prize, that there was no female Shakespeare, etc. I think her arguments contain a lot of social context: with the ability for opportunities, you can get recognition, or a sense of normalcy.

Now comes Freud. For him, the figure Wilhelm Fliess was extremely important because Freud communicated all the time with him, even though he was not working in the psychology field *per se*. So, again, there is a strong relationship.

Einstein is another example. There was a big controversy about his first wife, Mileva Maric, who is actually credited now for having worked out the mathematics of special relativity, and who was a joint author when they submitted the paper. Einstein then became interested in another woman, and all sorts of things happened. Also, given the atmosphere of the time in Vienna, which was very much opposed to giving opportunities to women, she was completely pushed aside.

As these examples show, social relation is very important in managing research. Nothing will happen if a genius is put in a room alone. The genius needs an audience (and also novel critical customers) in order to drive him or her forward to challenges. So, the audience is important. Professor Rochat talked about audience awareness: to communicate your idea, you need an audience to help you formulate the idea.

The issue of complementarity, mentioned by Professor Poeppel, is also important. Nobody is good at everything. Apparently, Mileva, Einstein's first wife, corrected his papers on mathematics and worked out some of the basic points. So, there was a complementarity between the two. Without her help, Einstein could not have achieved anything.

There is also the issue of motivation, which is the point Dr Mogi made. If you were left on an island alone, you would not have started scientific investigations.

Another concept is dialog and cooperation. Louis Pasteur had many assistants. Some of these people actually claimed to be the ones who made some of the major inventions; whereas Pasteur likened them to Eiffel, who claimed the design of the Eiffel Tower although somebody else actually did the design. There are many stories like this one.

So, the social factor or process is important in building up first peer recognition, and then public recognition. There are many people in science who deliberately manipulate the network of peers, including Einstein (who destroyed the document so that his wife would not be recognized). I can also think of Watson and Crick. If they had not obtained Rosalind Franklin's results, then they would not have arrived at their own results. Franklin's results were taken by Wilkins, and were shown to Watson and Crick without her knowing.

So, these people are famous for two things: getting the idea from others, and then not giving the proper credit afterwards. This is the "dark side" of science, if you wish. Some people are so heavily involved in marketing their own image, yet this is the strategy that to some extent works. Rosalind Franklin, for example, had not yet published her work, because she was not yet completely sure of her experiments. Then, all of a sudden, somebody else claimed to have found everything.

Children Play Together to be Creative

This is just to say that the social aspect is very much important. Now, I am switching from the genius to the child, because Ken Mogi said that we should also reflect on that.

A book has been written about a very fascinating preschool in Reggio Emilia, Italy, and I highly recommend it. To illustrate the Reggio Emilia approach to child education, let's say that a group of children are just playing together, sitting on the floor where a paper-made white pig is; then, by accident, they start tearing it. There is no instruction to do that.

The important thing I want to point out here is that in cognitive science research, the child is viewed as an individual and experiments are done individually. However, a lot of interesting stuff actually happens because of the group, and so the dynamics or the interaction within the group is important. This is the case not only in the field of science, but also in the field of arts. The individual artist is supposed to do everything, but even in arts social interaction is crucial.

Let us suppose that a child is trying to put a pen inside the stem, and cannot manage to do it. But, there is another child who is watching this and who tries to do what the other one was doing by holding the pen differently. So, this is not really imitation because the other one cannot do it. The latter child finally manages to do it.

It is through such a combination of chance and role that this world has evolved. It is the seeing of somebody else trying to do something, thinking, "Oh, maybe I can do that!" and then finally showing the other one how it is done. So, I would say that creativity (and any kind of feeling, for that matter) is very much of this nature.

So, if you look at the individual and argue that the individual is the center of the universe through which everything happens, then I think we are missing out the crucial ingredients of creativity. This is the first question I would like to raise: how do collaborative situations lead to creativity and learning?

Children and Representations

We now come to the second theme. It is about the importance of representations. By representations, I mean external representations, so I am not talking about brain-related or computer representations. Examples of external representations include drawings, languages, gestures, or pretended play.

Does anybody have an idea of what the following picture might represent?

• • •

Audience: Maybe inside a supermarket.

Steels: Supermarket. Any other ideas?

Audience: Someone is doing the laundry.

Steels: Laundry. Did somebody say a garden?

Audience: Yeah, that's it.

Steels: Actually, this is a bus [laugh].

• • •

It is worthwhile looking at this picture because it gives us the essential features of representations. First of all, there is the child, and there is the bus. Now, what is important about the representation of the bus to be an artistic drawing is the expression of meanings. There are various kinds of meanings, which are important for interacting with the world.

Notice that this picture is pretty big. The whole page is filled with the bus; you cannot make it any bigger. You can also see a conductor, so this is a British bus, which has many windows on top. So, you can see the windows, the driver, and the conductor who is going around to check the tickets, with his arm outstretched asking for the tickets. Another thing about the bus is that it has many windows and many wheels, so you see many wheels and many windows in the picture.

So, there is actually a grammar here. The picture is trying to express not just one wheel, but many. This is amazing. Also interesting is that you have different perspectives at the same time: there is a side perspective of the bus, but you can also see an overview of the window of the bus from the top. I could go on and on about the picture, but this is really what representation is all about. It is the expression of meaning; and we call it "creative" because this child, just an ordinary child, invented all these conventions, all these ways of expression.

The picture of the bus is not just a representation, but is also a story. Again, there are social interactions happening here. There is a kind of unity that goes beyond just one child acting alone.

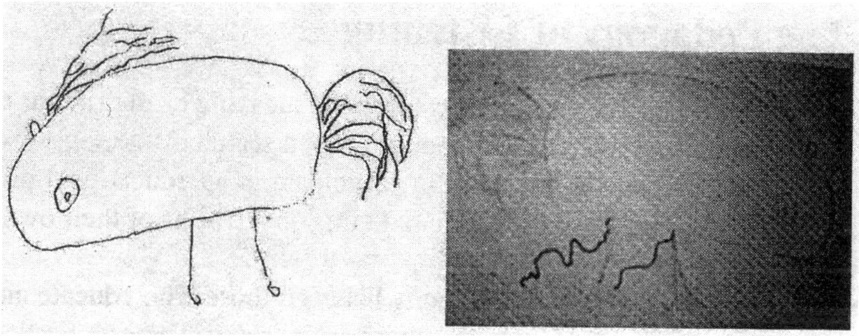

The picture above, shown by Carla Rinaldi,[a] is another example of creativity. The child, named Elisa, tries to draw a horse. Horses have four legs, but this one has only two legs. Elisa probably recognizes 3D images. You cannot just add the two legs here, as that would be a silly thing to do. So, how you deal with this problem? This is, of course, a problem that many painters have thought about. However, this child basically turns the page around and draws two legs on the other side. So, if you hold the paper up, it gives you the 3D feeling. This is fantastic, especially since the picture was drawn by a kid aged just 3 years 2 months. Thus, the conclusion, at least to me, is that representation can be extremely important at that age.

So, I think it is true representation-making that makes us human. Secondly, creativity shows up in all of us. Every child, not just the genius, is creative and has the power to create representations. Of course, after a

[a] Rinaldi C, *In Dialogue with Reggio Emila*, Routledge, London, 2006.

while, norms are imposed, such as the norm of having to be realistic. Many children stop drawing at 10 or 12 years old, partly because they cannot draw in a realistic way. In other words, a realistic drawing is different from the expression of meaning.

Representation and Language

The making of representations happens all the time in languages as well. Even though it is partly conventionalized in that there is a shared norm, we still invent part of our language every time we talk. So, for me, in contrast with the views of a lot of linguists like Chomsky, language is not a completely rule-governed, static system; rather, it is a system we all create together. So, every time you speak, you create new meanings or you may shift the meaning of the words.

In Nicaragua, at the end of the 1970s, the Sandinista took over the country. Now, before the revolution, deaf children were considered mentally retarded and were not educated. So, they were left in the care of their parents, who thought that they were failures and that nothing could be done for them. However, after the Sandinistas came to power, these children were brought together to get them to school. Now, the teachers in school did not know any sign language, and so they also thought that these children were incapable of actually learning sign language. But, what happened was that these children started to invent their own language in a very short period of time.

So, just the fact that they came together was enough to trigger the very fast creation of this new language, which was new in the sense that it was a sign language of their own invention, unlike American sign language or French sign language. By the way, French and American sign languages are very similar, whereas British sign language is very different from American sign language. So, those who understand French sign language can understand American sign language, but not British sign language, for instance.

People who learn sign language start with iconic representations, but then these representations become abstract; just like a natural language, where you start drawing from your experience in the world. Just as the word "back" becomes used by analogy (e.g. "in the back of a car" or "let's go back") and starts to propagate in all directions, the same goes with sign language.

The particularly interesting thing here is the use of grammar. The grammar in this case is expressed through the roles of objects in actions, like who is tapping whom. So, these gestures start with tapping and then are used by analogy in other situations, and eventually become like prepositions. This is the real-world example, fairly recent of actual phenomena like that where the new language was observed. Of course, if there is a norm, then you try to use the norm; but still, within the norm, each of us can be very creative.

Dialog by Editors

Summing Up: Dr Tokoro,* Dr Mogi, and Dr Sasaki†

1

Tokoro: Let's start with Professor Snyder's notion of "mindsets." As we become grown-ups, we learn from our past experiences, acquire knowledge, and — without knowing it — become accustomed to think in a routine manner. This tendency, of course, is not a bad thing. Having a mindset enables us to live socially and effectively.

So, what is creativity? Professor Snyder seems to propose that creativity lies in breaking the pattern. I think this is his main point.

I am impressed because he is a man of his words. He actually built a device that emits a very weak electromagnetic pulse (one to two hertz), thus affecting the left temporal lobe.

Professor Snyder came up with this idea when he was observing a patient who had damage to the left temporal lobe. He hypothesizes that once we get rid of the mindset, a thing that was somehow hidden from us comes directly into our consciousness. I think this idea is ingenious.

---

* Mario Tokoro is the cofounder and president of Sony Computer Science Laboratories, Inc., Japan; and senior vice president, corporate executive, Sony Corp., Japan.

† Takahiro Sasaki is a researcher at Sony Computer Science Laboratories, Inc., Japan.

He concludes that creativity has something to do with shutting off the ordinary way of looking at things, to see things in a new light.

Mogi: Professor Snyder's hypothesis is very interesting. At the same time, it is rather controversial. He tries to explain the savant syndrome: by suppressing part of the left temporal lobe, the right hemisphere becomes free and active.

One of the highlights of this conference was a difference of opinion showed by Allan Snyder and Luc Steels. Allan believes that a literal way of seeing the world, which is a typical trait of pictures drawn by savants, leads to genius. Luc, on the other hand, claims that the true nature of human cognition lies in children's somewhat clumsy drawings.

I think the ability Allan is trying to release by stimulating the left temporal brain is rather mechanical, i.e. seeing the world as it is, literally, like counting dates on a calendar. Clumsy drawings show the opposite extreme of human cognitive ability. I believe true geniuses possess both of these abilities.

Tokoro: The drawing of the horse that Allan cites in his talk indeed looks like the work of a genius. But, was that horse drawn by observing a live horse or by seeing a photograph? I think that makes a big difference for the value of the drawing.

Nobody seems to be able to draw a picture like that, simply because the horse is moving so fast. If it can be done, I am amazed.

Mogi: I understand that Nadia (the autistic girl) saw a galloping horse and drew the picture afterwards from her recollection.

Savant children are said to possess that kind of ability. You fly over the city of London and see a building. When you get off the helicopter, you draw a picture, and the number of windows exactly matches that of the real building.

Tokoro: So it *is* amazing.

When talking about creativity, I think we need to have our own perspectives. I mean, just being literal is not enough. When Allan says "mindsets," does he mean mindsets common to many people or individual mindsets? Anyway, I believe the most important thing when being creative is to create one's own point of view.

Mogi: I quite agree with you there. In general, savant children are said to be not really good at finding new meanings from the world around them. For example, they may remember a conversation word by word for 10 or 20 years; however, this kind of accurate memory is often accompanied by the failure to learn from experience. To learn from experience, you need to somehow draw meanings from the episodic memory that you gather through your interactions with people. Savant children are not particularly good at this point. There is a trade-off here.

As I understand it, in this trade-off, it is not enough to have one mindset or another. A genius is someone who has some traits of both mindsets at the same time.

2

Tokoro: Let's move on to Professor Poeppel's talk. According to him, human brains cannot tell which signal occurred first when the interval of two consecutive signals is less than 30 milliseconds. This is an experimental fact. It is quite interesting.

Then, there is the 3-second rule, which states that the human brain can only remember or perceive events that last less than 3 seconds as a single "pack" of meaning. Phrases of poems, music motifs, and even Japanese Noh plays have this 3-second unit in common.

This can be thought of as an "atom" of perception, i.e. the smallest unit of perception. This should come from the brain's physical structure. Professor Poeppel discovered that fact. It is intriguing.

The activity of learning or thinking is also bound by 3 seconds. Three seconds is the key to all human cognition.

Mogi: I think Professor Poeppel's points are indeed very interesting.

The brain can be thought of as a superparallel computing system. In order to synthesize miscellaneous elements of cognition, temporal "book-keeping" plays a central role. This book-keeping process seems to occur not only when receiving sensory inputs, but also when the brain acts on its own.

The problem of temporal book-keeping emerges when we begin to act and something happens as a consequence: there is a loop between our action and our senses. That is how we come to an understanding of the world, and Professor Poeppel is a pioneer in this field.

The integration of sensorimotor integration in terms of time is an important research issue. It is related to the concept of agency. The problem is interesting from the point of view of the synthesis of parallel processing. In ordinary terms, the problem of timing has been said to have something to do with music or art (especially in relation with music).

Einstein loved Mozart and played tunes with his violin. Why is music, where the timing is so important, related to creativity? We all know from experience that timing is important in creative acts. So, Professor Poeppel's hypothesis has many implications for creativity.

Tokoro: Let me point out an interesting coincidence: the TV screen is refreshed every 30 milliseconds!

Mogi: Yes, these facts are clearly related. When considering brain function, time makes the whole story interesting. Time is relatively easy to study because it is a one-dimensional parameter. The study of emotion, for example, becomes very interesting when time is taken into consideration.

Tokoro: Three seconds correspond with short-term memory. What is its relationship with long-term memory?

Mogi: The encoding and retrieval of long-term memory is closely related to the emotional system. The information first goes to the hippocampus, and then gets fed into the cortex. You can tell whether the information will be stored (as long-term memory) or not by looking at the activity of the hippocampus 3 seconds after receiving the information.

So, 3 seconds is the time required to make a (unconscious) decision whether to store a short-term memory into long-term memory or not.

Tokoro: When you say emotional system, I think that has something to do with dopamine, for example. When you have an "Aha!" experience, how long does it take for dopamine to be released?

Mogi: That is not yet known, especially in relation to the memory circuit. People, however, never forget events like September 11, for example. It is certain that an emotional system is responsible for deciding what to remember and what not to remember. The important thing is that it takes about 3 seconds.

Tokoro: And when it vanishes?

Mogi: For example, when we experience something very strong and shocking, let's say that event lasts for 1 second, after which we do not remember what happens. We remember a whole temporal sequence — be it 1 minute or 10 minutes — as an episode.

The temporal information processing is not well understood.

Tokoro: Now, what do all these have to do with creativity? Professor Poeppel says that creativity is based on both perceptual and conceptual processes. He also says that creativity is related to expression.

Mogi: I think 30 milliseconds is the unit of time in the perceptual domain. As for conceptual thinking, since the working memory is at work, in a sense, it would be related to 3 seconds.

Tokoro: What happens in the cache memory, that is, near 3 seconds?

Sasaki: I think Professor Poeppel is saying that the conceptual connection comes in when you widen and connect the 3-second window, when the time flows longer than 3 seconds to get the meaning. My impression is that both 30 milliseconds and 3 seconds amount to Professor Snyder's "literal perception," and the chain of 3-second windows corresponds to Professor Snyder's "mindsets".

Mogi: Well, the longer-than-3-second time flow is accounted for in the conceptual process. I think that when a concept is called into the working memory, the time unit is approximately 3 seconds. For example, human beings cannot glance from a bird's-eye view for 1 whole hour. So, we bind together the 1-hour event into a concept; this concept in turn is utilized by the working memory, which has a duration of 3 seconds.

Tokoro: Connecting a sequence of 3 seconds within the rule of 3 seconds. That's very interesting.

3

Tokoro: Dr Kitano's story is a bit different in its context. He talks about robustness and fragility. Interesting phenomena occur on the border of stability and instability. The examples he cites, such as cancer and diabetes, are rather difficult to understand. I think the brain is also stable and unstable at the same time.

Sasaki: I think Dr Kitano is saying that fragility is inherent or inevitable once the system becomes robust.

Mogi: Yes, I think that is an important point. For example, a stone can be said to be robust, but it is not interesting because it doesn't change. Robustness in a biological system is closely tied with fragility.

The brain is no exception. You can build a very stable cognitive system, but that would not be creative at all; that would be similar to a fixed software on a computer.

A creative information processing system should include fragility, and it should evolve in a robust manner. In that sense, the brain's information system and the biological metabolism have something in common.

We know that geniuses tend to be "marginal." A genius walks on a thin thread between fragility and robustness.

Tokoro: Dr Kitano cites an example of the stealth airplane. You can build an airplane that the enemy cannot detect or shoot down, but a weak point eventually shows up: a human pilot cannot fly the airplane anymore. It is amazing that such a product exists on the border of stability and instability.

Mogi: I agree. Sometimes a person who makes trouble, be it a crime or some kind of mental problem, tends to be creative. We cannot have geniuses without having troublemakers.

That is the nature of the brain. We have to cope with these problems. I think creativity comes out of a conflict between robustness and fragility.

Tokoro: As for criminal activities, I don't think it is right to label a person good or bad based on what he or she has done. We might fall in a situation like that, too.

Mogi: Yayoi Kusama, an internationally renowned artist, for example, has been hospitalized in a mental clinic, and her works are excellent.

Tokoro: So, fragility and robustness, or stability and instability, have much to do with creativity.

4

Tokoro: Professor Rochat's talk was very stimulating and persuasive. What fascinates me are the rigid scientific foundations or data that support his hypotheses.

Two months is the first critical point, and 9 months is the second critical point; then, there is another evolution at around 18 months to 2 years of age. Babies less than 2 months old are still not very different from prenatal babies. But, after 2 months of age, babies begin to realize that their bodies belong to them; here, the word "proprioception" is used. After 9 months, babies become somewhat social.

Human beings are social animals. I think Professor Rochat is claiming that this social behavior is closely related with Professor Snyder's "mindsets."

Professor Rochat says that creativity emerges when we stop worrying about what other people think. Does this mean that we should try to cut ourselves off from the social environment? Creativity cannot be conceived of as some kind of isolate, solitary, autistic activity.

Mogi: Professor Rochat looks deep into the social structure. He thinks that peer pressure plays an important role when it comes to creativity. For instance, a kid is happy with a red patch on his/her forehead, as long as other people around him/her have the same patch. This kind of peer pressure returns when he/she becomes a teenager; teenagers face another crisis in human cognitive development. I remember experiencing very strong peer pressure when I was in junior high school.

Tokoro: That's the age when you have to be the same as others, and still show that you are different.

Mogi: Professor Rochat's talk has much to do with Professor Snyder's research. Savant children tend to lack social awareness and behavior. They don't look through an ordinary looking glass. They typically experience difficulty in verbal communication.

So, Professor Rochat's notions of "self as intimately coupled with society" and "self as cut off from society" are very close to Professor Snyder's "see the world with mindsets" and "see the world free of mindsets," respectively.

In that sense, I believe Professor Rochat's claim and Professor Snyder's claim are very similar. They both claim that creativity is the problem of how to be free from mindsets.

Sasaki: I was talking with Professor Rochat, and found that he does research on tattoos. I asked him, "Why does a developmental psychologist get interested in the tattoo business?" He told me, "Tattoos have something to do with cutting ties from the rest of the pack." In the same sense, people who wear earrings and even people who cut their wrists or commit suicide try to draw others' attention — they try to be different, too.

Professor Rochat said that teenagers tend to have tattoos or pierced ears. He also mentioned the Japanese *Yakuza* ("mafia").

Tokoro: That sounds interesting. They try to get out of society, but at the same time they try to build their own community. People who wear earrings or who have tattoos on their bodies realize each other on the street.

Sasaki: They are desperately trying to get out of the standard community. They need something like rings or tattoos. But, being creative may be somewhat different.

Tokoro: Maybe that is related to Dr Kitano's notion of fragility.

Another interesting point in Professor Rochat's discussion is his critical attitude towards Jean Piaget's ideas. Piaget paid attention to the functional aspects of intelligence. Professor Rochat, on the other hand, claims that even simple examples like pulling a blanket can be interpreted differently. He says that babies of 9 months of age are interested in their relationship with others; the behavior of these infants no longer stays in the context of "how to use tools."

Mogi: Professor Rochat is one of the pioneers who claimed that human intelligence is social and is conditioned according to a person's relationship with others. Nowadays, this concept is regarded as common sense in the cognitive science community.

Tokoro: Was the situation different in Piaget's age?

Mogi: I believe Piaget had the same things in mind, but the atmosphere was totally different back then.

Sasaki: What about Vigotsky?

Mogi: Yes, Vigotsky is being frequently cited these days.

Tokoro: To sum up, Professor Rochat says that self-consciousness is the barrier to overcome if you want to be creative. This point is very interesting. If you are too self-conscious, if you care too much about what others think of you, then you cannot be creative.

So, creative people don't give a damn what other people think. They might get you into a lot of trouble, though. Beware!

5

Tokoro: Let's move on to Professor Masataka. He was a first-rate entertainer. I loved his performance.

He talks about physical development and its relation with expressional ability. He also discusses the problem of learning words. When we learn certain words, the name of an object, for example, we need to understand the person showing the object to us. Who is this person? What is his/her intention? And so on.

Mogi: Very interesting presentation.

The other day, at a meeting of the Society for Neuroscience in the United States, I listened to a talk by Dr Patricia Kuhl. She told us that a child's relationship with his/her mother is very important. For example, infants learn a language far better when actually facing their "teachers" than when looking and listening to them on a TV monitor. They need to talk to their mothers in person.

Tokoro: Isn't that story mentioned in Professor Rochat's speech, too? When mothers and kids talk over the video monitor, there is a significant difference between the video replay mode and the real-time mode. A language is definitely something social.

Professor Masataka also talks about an experiment using roman numerals, a story about acquiring a new concept. I wonder if you can do multiplication using roman numerals. Is it possible?

Mogi: When you get used to it, maybe. We can do calculations using Chinese characters.

Tokoro: I think you can do addition, but can you multiply?

Mogi: I don't want to try [laugh].

Sasaki: I personally found if interesting that when we acquire a new language, not only the language-processing centers, but also the motor-related areas play a big role. We learn words by looking at the gestures of the person in front of us. That's very interesting. Language acquisition is not just verbal or mental, but also physical.

Tokoro: Could you explain Professor Masataka's hypothesis about Einstein?

Sasaki: Professor Masataka claims that Einstein became a genius not because his brain was perfect, but because he had a defect in Brodmann area 39. Einstein suffered from aphasia. His working memory was defect. Since his articulatory loop was not functioning well, his visuospatial scratching pad took over the job. Professor Masataka claims that this was why Einstein became very good at geometry and math.

Mogi: I think the idea that Einstein became a genius as he overcame his difficulties is fascinating.

Tokoro: So, what do we do? We don't have aphasia.

Mogi: There are many kinds of geniuses. Goethe, for example, was very linguistic!

6

Tokoro: Dr Mogi, in your talk, you first prove that there is no homunculus in the human brain. When you say that it is sometimes better to pretend that a homunculus exists, what does that mean?

Mogi: I think we can say that all cognition is metacognition. It is usually thought that once the self comes in, metacognition emerges as something at a higher level. My claim is that every cognitive process is necessarily metacognitive. "I" comes into play and the problem of identity emerges; then, there is the parallel processing and the synthesis process in general.

I think that we are unable to solve the problem of consciousness, nor can we grasp the essence of the information processing mechanism, including creativity.

Tokoro: So, breaking mindsets and creating one's own mindset amount to giving birth to another homunculus in your head?

Mogi: Exactly. That is my hypothesis. While I was learning Dr Kitano's concept of robustness, I came upon an interesting concept: heterarchy. Heterarchy is exactly what Dr Tokoro mentioned just now.

In the case of hierarchy, there is only one top; heterarchy, on the other hand, has several tops. Heterarchy is closely related with robustness. People in ecology talk about these things.

Of course, Dr Kitano is an expert on robustness, so he talks about robustness, but many people actually discuss robustness in disguise. These people use the concept of heterarchy instead.

Tokoro: What happens to the old homunculus when a new one emerges? Is it vaporized?

Mogi: No, they just get switched.

Sasaki: Usually, healthy people synthesize many homunculi, I presume. What is the difference between ordinary state and multiple personality disorder?

Mogi: One key is the memory segmentation. A part of the right frontal lobe becomes active and suppresses the memory related to an experience of abuse in one's childhood, for example, and builds a fake memory. In order to be consistent, if you suppress abuse memory, you have to build another life history to replace it. This is how multiple personality disorder is believed to be generated.

A similar example is Capgras' syndrome. The brain's emotional system makes up a fake memory. This phenomenon is well known. The same thing happens to a pathological liar.

So, when you talk about the homunculus, you imagine it to be in the frontal lobe area 46 or the prefrontal cortex. However, the fact of the matter is that your emotional system is playing a central role behind the scenes.

The emotional system, such as the basal ganglia, is intimately related to the frontal lobe. When you talk about the emotional system, you often think of the limbic system; but, the lower part of the frontal lobe (in particular, the orbitofrontal cortex) is also related to this processing of emotion.

Tokoro: I would like to relate your story to Professor Poeppel's hypothesis. We often come up with seemingly good ideas within 3 seconds ("This is splendid!"), only to realize afterwards that we were wrong ("No! This condition doesn't fit.").

I think that it is not a simple matter to synthesize something into a homunculus. I mean this not only in terms of the mechanism, but we also do a lot in the process of being creative.

Mogi: Thomas Edison used to say, "Genius is 1% inspiration and 99% perspiration." I think he is quite right. Inspiration comes in an instant, but you have to spend 99% of the time making it into something worthwhile.

Tokoro: People say such things in science, but I think the same holds true in art. Painters usually draw hundreds of pictures before completing one final picture. They sometimes fail tens of hundreds of times to produce one color. They know what they want in a second, but they cannot make it real. I think art as well as science and engineering all have these processes inside the brain in common.

Mogi: There are many people who have the 1% part, but not the 99% part, and thus cannot succeed.

Tokoro: How about TV? TV is social. Suppose you watch too much TV. You may have a good idea within 3 seconds, but you cannot develop it into something real because you watch too much TV. TV games may be worse.

Mogi: I don't think TV monitors *per se* have any bad influence on the human brain. However, I do think that there are some important elements that are lacking from today's TV or TV games.

When kids play ordinary games (not TV games), they must set the rules and negotiate and communicate with each other. They have to do this before they start playing the game. If a kid is weak, they make rules that are slightly advantageous for that particular kid (i.e. a handicap).

This process is very important when being creative. The problem with TV games lies in the fact that the rules are already given. That's why TV games are not particularly good in nursing creativity.

Tokoro: Kids deliberately include fragility. It's not fun if the strong kid always wins the game. So, they modify the rules to give a chance to other kids.

Mogi: Yes, that is right. They produce risks on their own.

Tokoro: In the case of TV games, they say it's stage 1, then stage 2, and so on. The machine develops the rules for us. We have no say in modifying the rules. It's a problem of skills, not a problem of creativity.

Mogi: When we observe kids play, they control their risk levels by themselves. For instance, they may ask themselves, "Can I jump from this height?" This is a very important point, even as a research agenda in brain science.

Tokoro: Dr Mogi's talk was rather difficult to understand, but now I think I understand well. You are saying that the emotional system is behind our creation of many things, and that creativity is a reaction to the uncertain world.

Mogi: That is the biggest agenda of our brain research team.

7

Tokoro: Dr Steels' talk is unique. Usually, when you talk about creativity, the general assumption is that you are the sole source of creativity. What Dr Steel claims, on the other hand, is that it is through the interaction with others that we can be creative. He sheds light on these aspects of creativity.

It is said that being social and having a fixed mindset are negative properties. Dr Steels thinks that it is more important for a group of people or for society as a whole to be creative. Kids' play, as Dr Mogi mentioned, where kids set their own rules, is not a personal creativity, but a social creativity.

From an individual point of view, a mindset is considered as something negative. You can, however, still be socially creative — with a new mindset of the group. We came to know this new possibility in the workshop.

Unlike Professor Snyder, Dr Steels claims that realistic drawings *per se* are not creative. He claims that the expression of meaning is creative. So, the picture of a bus by a 2-year old kid is creative because it has meaning in it, just like a Picasso painting.

The concepts of group creativity and expression of meaning by representation are both important. Some people are good at drawing pictures alone, while others are good at being creative as a group.

Mogi: As I mentioned earlier, I think that one of the highlights of this workshop was the contrast between Professor Snyder and Dr Steels.

The expression of meaning is a focus of attention these days. Suppose a child draws two pictures: one is a lollypop candy and the other is a balloon. When we ask which is which, the child answers, "This one is a lollypop and that one is a balloon." This is in spite of the fact that the two pictures look exactly the same!

An expression of a picture is not just some objective physical property. It includes intentionality. Intentionality *is* the meaning. Dr Steels' problem leads us to the origin of letters and symbols, evolutionarily speaking.

Symbols and letters emerge out of a society through the interaction with others. In this sense, Dr Steels' claim naturally came out from his research on the social origin of language or the Talking Heads experiment.

Tokoro: The story of the Nicaraguan sign language is suggestive. The children created their own tools of communication. There must have been a leader, and somebody must have suggested to use certain signs. Together, they built a language; they even invented grammar.

Dr Steels' research hovered around the origin of language from the point of view of evolution. Now, it seems he has gone a step forward. A language is an expression of the intention to communicate.

In fact, nothing comes out of the blue. Every product, even science, is based on the invisible effort of many people. Maybe we should reconsider patents and copyrights from the point of view of creativity.

Mogi: Darwin's theory of evolution is a good example. Darwin said it, Wallace said it, and other people said it in the past before them. Many people came close to the theory of evolution. This is not to diminish Darwin's originality, but it is important to shed light on the collaboration of people that led to the synthesis of the theory of evolution.

Tokoro: And Darwin did the 99% part, in the sense of Edison. He traveled around the world.

Mogi: Where lies creativity? Who has the rights to creativity? Creativity in a person might be a myth.

Tokoro: In general, great inventors or researchers are people who do the 99% part. One hundred people who do 1% of a job will not lead to success. We need someone who does the rest, i.e. the 99%.

Mogi: I heard someone say, "The Nobel Committee is nothing but an 'Aha!' detector." The committee does research on who experienced the first "Aha!" in the world. For example, Dr Koichi Tanaka would never have had the chance to receive the Nobel Prize had it not been for the intensive research done by his committee. The Nobel Committee is only interested in the first 1%.

Tokoro: Now, let's sum up. What can we do to be creative, especially in education at school and at home?

Mogi: One, break mindsets. Two, interaction with others. Three, the 99% part or the perspiration. I think that's about it.

Index